Physics Fixing

SANDEEP CHAVAN

Published by SANDEEP CHAVAN, 2025.

PHYSICS FIXING

First edition. February 10, 2025.

ISBN: 979-8227458322

Written by SANDEEP CHAVAN.

Table of Contents

Prologue 1

1: The Fixing of Physics 6

2: The Space-Time Fantasy 54

3: The Quantum Cover-Up 75

4: The Nuclear Force Myth 95

5: The Standard Model Disaster | A Frankenstein of Fixes 115

6: The Gravity Illusion 134

7: The Time Illusion 154

8: Einstein vs. Tesla 173

9: Structured Vacuum Energy 196

10: The Atomic Truth 214

11: Gravity as an Energy Gradient 232

12: Redefining Light 252

13: The Future of Science 273

Epilogue 296

The Road Ahead 301

Rewriting the Foundations 304

Experimental Evidence & Future Tests 308

FAQs & Common Objections 312

Open Problems & Future Research Directions 319

Acknowledgments 324

Further Reading 326

About the Author 330

Disclaimer 332

To **physicists, past and present,** who have shaped our understanding of the universe—whether through groundbreaking discoveries or by daring to challenge established norms.

To **independent researchers** who continue to explore the unanswered questions of physics, often without institutional support, driven only by their relentless pursuit of truth.

To **physics lovers and deep thinkers**, whose curiosity knows no bounds and who refuse to accept contradictions as final answers.

And to **my dear students**, the future minds of science, who will question, explore, and redefine the very foundations of physics.

This book is for all of you—May the search for truth never stop.

Sandeep Chavan

"Truth in science is not about preserving traditions, but about daring to question them. The universe does not bend to equations—it reveals itself to those who seek beyond the boundaries of convention."

— Sandeep Chavan

Preface

For centuries, physics has been regarded as the ultimate pursuit of truth—a discipline that seeks to unravel the mysteries of the universe. From Newton's laws to Einstein's relativity and quantum mechanics, our understanding of reality has evolved through groundbreaking discoveries. Yet, despite its successes, modern physics stands on a fragile foundation. Instead of offering a unified, coherent picture of the universe, it has become a collection of mathematical patches, contradictions, and incomplete theories.

This book, *Physics Fixing: A Bold Truth for the Future of Science*, is an attempt to question everything we take for granted about physics. It does not aim to refine existing models—it aims to rebuild physics from the ground up by exposing the flaws in mainstream theories and proposing a structured energy-based framework as a more logical, unified, and fundamental explanation of reality.

Throughout my years of research and critical inquiry, I have encountered several deep contradictions in modern physics:

- Gravity is treated as space-time curvature, yet space itself is never defined as a real entity.
- Light is said to behave as both a wave and a particle, yet no one can explain why it "chooses" one form over the other.
- Quantum mechanics is built on probability, yet it offers no physical mechanism for how or why wave functions collapse.
- The Standard Model introduces countless particles and force carriers, but instead of simplifying physics, it makes it more complicated.
- Time is considered a fourth dimension, yet it has no independent physical existence.

The more I explored these issues, the more I realized that modern physics is not a complete theory of nature—it is a patchwork of fixes that attempt to force mathematical predictions to align with observations. Instead of revisiting fundamental assumptions, physics has chosen to add complexity to equations rather than question their validity.

The Purpose of This Book

This book is not just a critique of modern physics—it is a call for a scientific revolution. It aims to:

- Expose the flaws in current theories, including relativity, quantum mechanics, and the Standard Model.
- Propose a new structured energy model, where forces, gravity, and even time emerge as effects of energy interactions.
- Encourage open inquiry and critical thinking, challenging physicists, researchers, and students to seek fundamental truth rather than mathematical convenience.

The structured energy model introduced in this book eliminates the need for force carriers, wave-particle duality, and space-time distortions. It offers a more natural explanation for gravity, electromagnetism, atomic structure, and the nature of light—one that does not rely on mathematical tricks but instead aligns with physical intuition and experimental reality.

Who Should Read This Book?

This book is written for physicists, researchers, students, engineers, and anyone with a deep curiosity about the nature of reality. You do not need to be a professional scientist to grasp the key ideas presented here—you only need to have a willingness to question conventional wisdom and explore new possibilities.

I encourage you to approach this book with an open mind. Some of the ideas presented here may seem unconventional, but all great scientific revolutions began with ideas that challenged the establishment. Newton challenged Aristotle. Einstein challenged Newton. Now, it is time to challenge Einstein, quantum mechanics, and the Standard Model itself.

The future of physics depends on whether we choose to cling to outdated models or seek a deeper, more fundamental understanding of reality.

Are you ready to question everything?

Let the journey begin.

Sandeep Chavan

Prologue

When Einstein Visited My Dream

I still remember that evening vividly. The lecture had gone well. I had just finished explaining the photoelectric effect to my students—a topic that made Einstein famous, earning him the Nobel Prize in 1921. I broke it down step by step, showing them how light behaves as discrete packets of energy called photons, knocking electrons out of a metal surface.

The students seemed engaged, nodding along. But just as I wrapped up, one of them—grinning mischievously—raised his hand.

"Sir, has anybody ever seen these photons? Or electrons? Or space-time? Or are we just supposed to believe in them because textbooks say so?"

The class erupted in laughter. A few others joined in:

"Yeah, Sir, has anyone ever caught a photon in their hand?"

"Or curved space-time in a bottle?"

"Or at least seen an electron, other than in a diagram?"

I smiled, but inside, something shifted. They were joking, but their questions cut deeper than they realized.

Had we, as scientists, spent a century accepting things we couldn't truly observe—things we couldn't even measure directly? Were photons, space-time curvature, and quantum particles just mathematical constructs? Were we blindly fixing problems in physics instead of truly solving them?

That night, as I drifted into sleep, my mind was still turning over these thoughts. And then—Einstein appeared in my dream.

A Conversation with Einstein

He stood there, his signature wild hair slightly more tamed than usual, his piercing eyes filled with something between amusement and curiosity.

"So, they asked you whether anyone has actually seen a photon?" he said, a smirk playing on his lips.

I nodded.

"And what did you tell them?"

"I told them that photons are real, that we know they exist because we observe their effects. The photoelectric effect proves it."

He chuckled. "Ah, yes. The photoelectric effect. My greatest trick!"

I frowned. "Trick?"

"Yes. The photon was my escape. I had to explain how energy was behaving, so I gave it a name: the photon. But I never actually proved that light is made of tiny particles. I just made the math work."

I felt a strange unease. Was Einstein admitting that even he wasn't fully convinced?

I pressed further. "But quantum mechanics, relativity—your work changed everything. Space-time, wave-particle duality, the speed of light being absolute—these ideas form the foundation of modern physics."

His face darkened. "Yes, and that's exactly the problem."

"What do you mean?"

He sighed. "Do you know what physics has become? A series of patches. Fix after fix after fix. Every time we couldn't explain something, we invented a new particle. When space didn't behave the way we expected, we curved it. When gravity didn't fit, we added dark matter. And when quantum mechanics didn't align with reality, we said it's because reality itself is uncertain."

"But the math works!" I protested.

"The math always works. That doesn't mean the universe does."

Silence fell between us.

Then, with a small smile, he looked directly at me. "You already know this, don't you? That's why your students' questions bothered you."

I swallowed. "Maybe."

"Then fix it," he said simply.

Why This Book Exists

I woke up with a start, my mind racing. Einstein was right.

For too long, physics had been a series of temporary fixes—a patchwork of theories that worked mathematically but didn't truly explain reality.

- We curved space-time instead of understanding energy gradients.
- We invented photons instead of exploring structured energy fields.
- We called quantum behavior "random" instead of asking what was really happening beneath it.

This book is not just a critique of modern physics—it is an exposure of how science has become a collection of patched-up ideas.

It will first break down the myths we've accepted as truth—and then rebuild physics on a foundation that actually makes sense.

Einstein may not have admitted his doubts openly, but he knew the truth: we've been fixing physics instead of understanding it.

This book is about ending the cycle of patches and finally finding the real answers.

Now, let's begin.

Part 1: The Inheritance of Fixes – How Science Became a Patchwork of Contradictions

1: The Fixing of Physics

How Science Became a Collection of Patches

Science was never supposed to be a series of quick fixes. It was meant to be a relentless pursuit of truth—an uncompromising quest to understand the universe as it is, not as we want it to be. Yet, somewhere along the way, we lost sight of this purpose. We stopped searching for answers and started patching over problems.

- A contradiction appears? Invent a new particle.
- An equation doesn't balance? Curve space-time.
- A force doesn't work? Create another one.
- A theory fails? Don't abandon it—just modify it until the math fits.

And so, physics became a tangled web of assumptions, mathematical tricks, and convenient explanations—a Frankenstein's monster of models that work just enough to be useful but never enough to reveal the fundamental truth. The worst part? Each generation of physicists inherited these patches and added more, never questioning the foundations.

This book is not here to respect tradition or to worship the past. It is here to tear apart the illusions, to expose the cracks in the very foundation of modern physics, and to show how science became a maze of contradictions held together by fixes, not understanding.

This is where we begin—not by looking forward, but by going back, uncovering how we got lost, and asking the one question that modern physics fears the most: What if we've been wrong all along?

1.1 The Moment We Stopped Asking "Why"

The greatest minds in history were not fixers; they were questioners. They didn't settle for explanations that simply "worked" on paper—they wanted to know why things happened. Newton didn't just describe gravity, he wanted to understand what caused it. Faraday didn't just experiment with electricity, he wanted to uncover its nature. Maxwell didn't just manipulate equations, he revealed an unseen order to electromagnetism.

But something changed.

There was a time when physics was driven by curiosity—by a desire to uncover the deep mechanisms of nature. Questions were fundamental, and no answer was sacred if it failed to explain reality. But at some point, the goal shifted. Instead of understanding nature, physics became an exercise in problem-solving.

The difference is subtle but profound. Problem-solving science does not seek truth—it seeks functionality. It does not ask, "Why does this work?" It only asks, "How can I make it work?"

This shift was the beginning of the end.

The Birth of the "Fix-It" Mentality

The moment physicists stopped asking "why," they started asking "how do we adjust the math to fit?"

Consider this: Newton's laws worked perfectly—until they didn't. When Mercury's orbit refused to follow Newtonian mechanics, instead of questioning the fundamental nature of gravity, physicists looked for patches. They first proposed hidden particles or additional celestial bodies affecting Mercury's motion. When that failed, they moved the goalpost—waiting for a mathematical fix. It took Einstein to introduce a deeper layer of reality with General Relativity, but even he didn't explain gravity—he just curved space to make the math fit.

The same pattern repeated again and again.

- Light didn't behave like a wave? Introduce the photon.
- Quantum mechanics made no logical sense? Accept probability as fundamental.
- The atom didn't hold together under classical physics? Invent the strong nuclear force.
- Gravity refused to fit quantum mechanics? Just call it an unsolved problem and move on.

Each time, physics took a shortcut instead of searching for a fundamental explanation. Each time, fixes replaced curiosity.

And before long, these fixes became permanent truths.

The Silent Shift: When Equations Became Sacred

There was a time when an equation was a tool to describe nature, not an unquestionable law. But the 20th century changed everything.

The rise of mathematical physics brought a dangerous belief:

"If the math works, don't ask questions."

This was the moment physics stopped being about reality.

Equations were no longer just descriptions of the universe; they became untouchable. If an equation predicted something nonsensical, physicists didn't challenge the equation—they assumed nature must somehow work that way.

- Wave-particle duality was absurd, but the math worked—so it must be real.
- Quantum randomness violated causality, but the equation predicted outcomes—so randomness became a fundamental truth.
- The Standard Model required over a dozen arbitrary parameters—so instead of questioning it, scientists just kept

tweaking the numbers.

Physics had become a religion of mathematics, where the question of "why?" was replaced by the command "shut up and calculate."

The Consequence: A Frankenstein's Monster of Fixes

The result? A patchwork of theories, stitched together with fixes and forced compromises.

- General Relativity describes gravity but fails in black holes and at the quantum scale.
- Quantum Mechanics works, but nobody can explain what's actually happening.
- The Standard Model classifies particles, but it's an artificial construct, not a true fundamental picture of reality.

And the deeper you look, the worse it gets. Every theory is held together by duct tape.

Instead of unification, we have multiple disconnected models—each valid in its own domain but completely incompatible with others. The ultimate proof that physics is broken is that it still cannot unify itself.

Why? Because physics stopped asking "why."

The Way Forward: Restoring the Lost Question

If physics is to be saved, it must return to its roots.

We must abandon the fixes, the patches, and the blind acceptance of contradictions. We must tear apart the assumptions that were built on convenience rather than truth. We must stop defending broken models and start questioning them—no matter how uncomfortable the answers might be.

We must, once again, ask the most dangerous question in science:

"Why?"

Because the moment we stopped asking that question, we stopped doing real physics.

1.2 The First Fix: When Physics Chose Convenience Over Reality

The greatest discoveries in physics were not born out of certainty—they were born out of crisis. Every major leap in science came when nature refused to behave the way we expected. But with each new discovery, scientists faced a choice:

- Dig deeper, challenge assumptions, and rebuild the foundation.
- Or apply a quick fix, force the equations to work, and move forward.

More often than not, they chose the fix.

The first great fix in physics didn't come from a lack of intelligence. It came from a lack of willingness to question the deeper nature of reality. And once that first fix was accepted, more followed—layer upon layer—until physics became what it is today: a collection of mathematical adjustments rather than a true understanding of the universe.

The Early Crisis: When Nature Defied Common Sense

Imagine you are an 18th-century scientist. The world makes sense. Newton's laws explain everything—from falling apples to planetary motion. The universe is like a giant clockwork machine, predictable and rational.

Then, problems start appearing.

- The orbit of Mercury doesn't match Newtonian predictions—a clear sign that something is missing.
- Light behaves like a wave in some cases and like a particle in others—an impossible contradiction.
- Electrical and magnetic fields are intangible but clearly real—defying Newton's laws.

Each of these problems threatened the stability of classical physics. Scientists had two options:

1. Admit that something fundamental was wrong and rethink everything.
2. Apply a fix and keep moving forward.

They chose the fix.

Newton, Maxwell, and Einstein—Revolutionaries or Fixers?

Let's look at the giants of physics—not just for what they discovered, but for what they left unanswered.

Newton's Laws – The Birth of the First Fix

Newton's force-based mechanics explained motion, but it completely ignored what caused gravity in the first place.

When asked how gravity works, Newton famously responded:

"I frame no hypotheses."

Translation: "I don't know, but the math works, so let's move on."

That was the first fix—accepting gravity as an unexplained force instead of asking if it was something deeper, like an energy effect.

Maxwell's Equations – The Invisible Fix

James Clerk Maxwell gave the world electromagnetism—a stunningly accurate theory that showed light was an electromagnetic wave. But there was a problem:

- If light was a wave, it should require a medium to travel through, just like sound waves need air or water waves need water.
- Scientists called this medium the ether—an invisible, all-pervading substance that supposedly filled space.
- But no experiment could detect it.

Instead of rethinking the nature of light, physics simply abandoned the ether altogether. They said, "Light doesn't need a medium. It just works."

Another fix.

Einstein's Relativity – The Most Elegant Fix of All

Einstein didn't just fix physics—he revolutionized it. But his genius was also his greatest limitation.

When he developed Special Relativity, he had to assume that the speed of light was constant. Why?

- Because Maxwell's equations said so.
- Because the Michelson-Morley experiment failed to detect ether.
- Because without that assumption, the equations broke down.

But instead of asking, "Is there a deeper structure to space that explains why light behaves this way?", Einstein simply declared that space and time must adjust themselves to keep the speed of light constant.

This was the most powerful fix in physics—but it was still a fix.

- Instead of explaining why light moves the way it does, relativity just reshaped space and time to make it work.
- Instead of discovering a structured vacuum energy that could explain gravity, Einstein curved space-time.

- Instead of treating gravity as an energy interaction, he made it a geometric effect.

And physics accepted it—because the math worked.

The First Great Fix: Ignoring the Structured Vacuum Energy

Here's the truth that physics ignored for over a century:

- Space is not empty.
- Vacuum is not "nothing."
- There is an underlying energy structure to reality that physicists chose to ignore.

In the early days of physics, scientists assumed that empty space contained a background energy—a structured vacuum that could explain why fields and forces worked the way they did.

But when experiments failed to detect this energy directly, it was erased from physics.

Instead of asking:

- "Is our method of detection flawed?"
- "Is this energy hidden in ways we don't yet understand?"
- "Should we redefine what 'empty' space actually means?"

Physicists took the easy way out. They declared space truly empty and forced equations to work around that assumption.

And that's when everything started going wrong.

- Without structured vacuum energy, physicists had to invent virtual particles to explain quantum effects.
- Without structured vacuum energy, gravity had to be explained as

space-time curvature, not an actual force.
- Without structured vacuum energy, the universe had to be filled with invisible dark matter and dark energy to explain missing mass.

Instead of admitting, "We might have misunderstood space itself," physics continued to apply more and more fixes.

And now, we have an entire model of physics built on top of a missing foundation.

The Moment That Changed Everything

The first fix—ignoring structured vacuum energy—set physics on a path where every future theory would require more and more patches.

This is why:

- The Standard Model is an unnatural mess of 17+ particles.
- Quantum mechanics and relativity contradict each other.
- Physics needs invisible forces, dark matter, and extra dimensions just to make equations work.

Physics never went back to fix its biggest mistake.

This book will.

We will go back to the root of the problem and ask the question that should have been asked over a century ago:

- What if space isn't empty?
- What if energy fields structure the universe at the deepest level?
- What if the first fix was the moment physics lost its way?

This is where we begin to undo the patches and rediscover the truth.

1.3 The Illusion of Scientific Progress

For centuries, humans have measured progress by what they build. A taller skyscraper, a faster car, a more powerful computer—each one a visible marker of advancement. Science is no different. The more we discover, the more we celebrate.

But here's the uncomfortable truth: what we call "scientific progress" in physics is often just a series of clever fixes for problems we never truly solved. The deeper we go, the more we find that every so-called breakthrough is nothing more than a patch to hold together an already broken system.

The illusion of progress is often stronger than progress itself.

When More Doesn't Mean Better

In the early days of science, a breakthrough meant understanding something fundamental. Newton didn't just describe motion—he gave us an entire framework for understanding forces. Maxwell didn't just write equations—he revealed the deep connection between electricity and magnetism. Einstein didn't just fix Newton's gravity—he replaced it with a new way to think about space and time.

But after Einstein, something changed.

- Instead of simplifying physics, we kept making it more complicated.
- Instead of finding theory that unifies, we built separate models that barely hold together.
- Instead of solving contradictions, we ignored them or patched them up with new assumptions.

The Particle Zoo – When Physics Became a Numbers Game

The clearest example of this illusion is the Standard Model of Particle Physics.

Once upon a time, physics had three basic particles—protons, neutrons, and electrons. Simple. Elegant. But then, experiments started showing strange results. Particles were behaving in ways that couldn't be explained. Instead of questioning the deeper structure of matter, physics did what it always does—it added more fixes.

- Can't explain proton behavior? Invent quarks.
- Need to hold quarks together? Create gluons.
- Need to explain weak nuclear interactions? Introduce W and Z bosons.
- Particles seem to have mass? Postulate the Higgs boson.
- Standard Model doesn't explain gravity? Ignore gravity.

And suddenly, physics went from three simple particles to over 17, with each new one existing only because the equations needed them to exist.

Is this progress—or just complicated bookkeeping?

Instead of asking:

- What if we're wrong about what particles are?
- What if mass is an effect of structured energy, not a separate particle?
- What if forces emerge naturally from energy gradients instead of force-carrying bosons?

Physics refused to question itself. It just kept stacking more patches on top of an unstable foundation.

The Standard Model is celebrated as one of the greatest achievements of modern science. But in reality, it's a desperate attempt to hold everything together without addressing the real problem: the missing foundation of physics.

Forces, Dimensions, and the Multiverse – How Far Will We Go to Avoid the Truth?

It's not just particles. The illusion of progress has extended into how we think about forces and even the structure of the universe itself.

- Gravity doesn't fit with quantum mechanics? Invent extra dimensions.
- The universe's expansion doesn't make sense? Postulate dark energy.
- General relativity breaks at singularities? Say "physics stops working" and move on.
- The Standard Model is incomplete? Invent supersymmetry to create even more particles.

At what point does adding more stop being progress?

Instead of searching for the missing truth about space, energy, and mass, physics has chosen to build a giant scaffolding of equations, each layer resting on another fix.

And the deeper you go, the clearer it becomes: we're not advancing—we're running in circles.

Nobel Prizes for Fixing Old Mistakes

Every year, the Nobel Prize in Physics celebrates groundbreaking discoveries. But if you look closely, you'll notice a trend:

- Einstein won for the photoelectric effect—because light didn't behave as waves, so he fixed it with photons.
- Dirac won for antimatter—because quantum mechanics needed symmetry, so he invented negative-energy particles.
- The Higgs boson was discovered—because the Standard Model equations required mass to have a source, so we added a field to

explain it.

The Nobel Prize doesn't reward discoveries of new fundamental truths—it rewards the best patches for old problems.

Imagine if medicine worked this way:

- Instead of curing a disease, doctors just kept prescribing more drugs to manage symptoms.
- Instead of understanding the root cause, they just kept introducing new medications.

That's what physics has become—a constant cycle of treating symptoms instead of curing the disease.

And the greatest irony? The bigger the patch, the more it's celebrated.

But a patched-up system is still broken.

The Real Measure of Scientific Progress

What if progress isn't about adding more?

What if real progress means removing the unnecessary and finding the simple, elegant truth underneath it all?

Physics used to do this:

- Newton gave us three laws to describe motion.
- Maxwell gave us four equations to describe electromagnetism.
- Einstein gave us one equation to describe relativity.

Yet today, the Standard Model has over 17 particles and multiple forces, and it still doesn't explain everything.

True progress doesn't come from adding more layers—it comes from peeling them away.

Breaking the Illusion

If physics wants to be saved, we must stop mistaking complexity for progress.

- We must question the foundation, not keep adding new fixes.
- We must rethink particles, forces, and space itself—not just tweak equations.
- We must dare to ask the dangerous question: What if everything we built was wrong?

Because the moment we stop believing in the illusion of scientific progress, we can finally start looking for the real truth.

And that's where the next section takes us.

1.4 The Religion of Mathematics – When Numbers Replaced Understanding

There was a time when physics was about understanding reality. Today, it is about defending equations.

Somewhere along the way, science stopped being a quest for truth and became a game of mathematical gymnastics. As long as the numbers fit, physicists no longer care if the theory makes sense. Contradictions are not seen as warnings but as puzzles to be patched. Impossible concepts are not questioned but rebranded as "quantum weirdness" or "spacetime curvature."

We no longer search for reality. We worship the math.

The Slow Death of Physical Intuition

The greatest physicists in history didn't just write equations—they built mental pictures of reality.

- Newton saw force as a tangible cause of motion.

- Maxwell visualized electromagnetic fields as real, structured entities.
- Einstein imagined light beams traveling through space to arrive at relativity.

They didn't just manipulate symbols on a page. They thought deeply about what the universe must be like.

But today, intuition is dead.

- Students are taught to accept the math without asking what it means.
- Quantum mechanics is treated as beyond human comprehension—so we just calculate probabilities.
- General relativity is described in complex tensors—but no one can truly explain what space-time is.

The deeper physics goes into abstraction, the further it drifts from reality.

Contradictions Don't Matter—As Long As the Math Works

A great scientist once said, "If a theory contradicts itself, it must be wrong."

Not anymore.

Today's physics embraces contradictions as long as they don't break the equations.

- **Wave-particle duality** – Light is both a wave and a particle, depending on how we observe it. Does this make sense? No. But the math works.
- **Quantum entanglement** – Particles affect each other instantly, faster than light. Is this logical? No. But the math works.
- **Virtual particles** – Particles that don't really exist pop in and out of existence. Is this physically real? No. But the math works.

- **Renormalization** – When an equation gives infinite answers, physicists erase the infinities and move on. Is this honest? No. But the math works.

Instead of questioning these absurdities, physicists simply accept them and continue calculating.

The Sacred Belief in Mathematical Consistency

Modern physics is built on one core belief:

"If the equations are consistent, the theory must be correct."

This belief has become so powerful that no one dares to challenge it. If a theory doesn't match reality, physicists don't question the math—they question reality itself.

- Quantum physics tells us the universe is random—so we assume randomness is real.
- General relativity predicts singularities—so we assume physics literally stops at black holes.
- Dark matter doesn't interact with anything—but we keep it because it makes the equations work.

We no longer search for what is true. We only search for mathematical consistency.

This is not science. This is religion.

How Mathematics Became a Shield Against Questioning

If you challenge modern physics, the response is always the same:

- "The math works, so it must be true."
- "You don't understand the equations, so you can't question them."

- "Reality is counterintuitive, so stop expecting it to make sense."

But this is not how science was built.

- Einstein didn't accept Newton's laws just because the math worked—he questioned them.
- Maxwell didn't accept classical mechanics just because it was useful—he expanded it.
- Planck didn't blindly trust existing physics—he introduced quantum theory.

The greatest breakthroughs in science came not from worshiping equations but from challenging the foundations.

Yet today, anyone who questions the Standard Model, relativity, or quantum mechanics is dismissed as a crank.

Why?

Because the modern priesthood of physics fears one thing above all else:

That someone might prove the entire foundation is flawed.

Why Scientists Fear Questioning the Foundations

If modern physics is truly correct, then questioning it should be welcomed.

But instead, questioning is silenced.

- **Physicists hesitate to challenge mainstream ideas**—because careers, funding, and reputations depend on defending the existing system.
- **Research that questions the Standard Model is rarely funded**—because it threatens the foundation of modern physics.
- **Young scientists are trained to calculate, not think**—because

deep questioning leads to difficult truths.

And so, physics has become a belief system. It operates not by curiosity, but by authority.

If a new theory contradicts the current models, it is dismissed instantly, no matter how logical it is.

The fear is simple:

If physics has been wrong for the past century, what does that say about the entire scientific establishment?

Instead of facing that question, scientists defend the old system at any cost.

Breaking the Religion of Mathematics

It's time to return physics to what it was meant to be.

- A search for understanding, not just mathematical consistency.
- A willingness to question everything, even if it threatens old models.
- A return to intuition, logic, and fundamental truths, not blind faith in equations.

Mathematics is a tool, not an answer. If the equations work but contradict reality, then the equations need to change—not reality.

- This book is not here to destroy physics. It is here to save it.
- But first, we must remove the mathematical blindfold that modern science has placed over itself.
- Because until we do, physics will remain a house of numbers, built on a crumbling foundation.

1.5 The Quantum Trap: When Scientists Stopped Looking for Reality

In the early days of physics, scientists sought to understand what was real. They wanted to know why things happened, what caused them, and what laws governed them. But with the rise of quantum mechanics, something changed.

Physicists stopped asking what reality is and started accepting whatever worked mathematically. If something was incomprehensible, it wasn't questioned—it was just accepted as a mystery of nature.

Nowhere is this more evident than in the birth of wave-particle duality, a concept that should never have existed but was embraced without hesitation.

The Birth of Wave-Particle Duality – A Fix, Not an Explanation

Once upon a time, light was believed to be a wave. Thomas Young's double-slit experiment (1801) showed that light interferes with itself, just like water waves. The evidence was clear—light was a wave, nothing more.

Then, in 1905, Einstein changed everything.

The photoelectric effect showed that light could eject electrons from a metal surface, but only if it arrived in discrete packets of energy. This was impossible to explain using the wave model.

So instead of questioning whether the nature of light was misunderstood, Einstein proposed a fix:

- Light must also behave like a particle in some situations.
- He called these energy packets photons—a concept that had no physical explanation but made the math work.
- Suddenly, light wasn't just a wave—it was both a wave and a particle.

This was the birth of wave-particle duality—not as a discovery, but as a forced contradiction to fix a problem.

Instead of finding the real nature of light, physics accepted an impossible paradox.

And once this door was opened, physics descended into madness.

When Probability Replaced Reality

If physics was willing to accept that light could be two contradictory things at once, it wasn't long before everything in the quantum world became paradoxical.

- Electrons didn't just orbit the nucleus—they existed everywhere at once until observed.
- Particles weren't solid objects—they were mathematical wavefunctions that collapsed into reality when measured.
- Reality wasn't absolute—it was a collection of probabilities, with no deeper explanation.

This was the trap.

The golden rule of science had always been: If a theory doesn't make logical sense, question it.

But with quantum mechanics, that rule was thrown away. Instead, physicists told themselves:

"If it makes no sense, it must be because reality itself is nonsensical."

And so, the quest for understanding ended.

Schrödinger's Cat – A Joke That Became a Foundation of Science

One physicist saw through the madness. His name was Erwin Schrödinger, and he wasn't impressed.

To mock the absurdity of quantum mechanics, he proposed a thought experiment that has since become one of the most famous (and misinterpreted) ideas in physics.

- Imagine a cat inside a box.
- Alongside the cat is a radioactive atom that has a 50% chance of decaying in an hour.
- If the atom decays, it triggers a device that releases poison, killing the cat.

According to quantum mechanics, before the box is opened, the cat is in a superposition of states—both alive and dead at the same time. Only when an observer looks inside does reality "collapse" into a definite outcome.

Schrödinger's point was simple:

- If quantum mechanics is correct, then reality doesn't exist until we observe it.
- This is absurd.

Yet, instead of seeing Schrödinger's cat as a critique of quantum theory, physicists took it seriously.

"Yes," they said, "this is how reality works."

And so, a joke became a foundation of modern physics.

The Problem with Quantum Mechanics

The problem is not that quantum mechanics is wrong—it clearly works. The problem is that it abandoned the search for an underlying reality.

- It stopped looking for causes and started accepting statistical outcomes.
- It stopped searching for what electrons, photons, and atoms actually are—and replaced them with mathematical wavefunctions.

- It turned reality into an equation, rather than something that could be understood.

Instead of asking:

- What is an electron, really?
- What is happening beneath the wavefunction?
- Why does measurement collapse probability into reality?

Physics decided:

"It doesn't matter. The math works. Move on."

This is the great quantum trap. We have built an entire branch of physics that predicts outcomes without explaining the cause. We no longer search for what reality is. We only calculate what we expect to see.

Escaping the Trap – The Need for a Deeper Theory

If physics wants to move forward, it must stop treating contradictions as truth and start looking for what's really happening beneath quantum mechanics.

- **Electrons, photons, and atoms are not random objects**—they follow deeper rules we don't yet understand.
- **Superposition isn't magic**—it is the result of a hidden structure that physics has ignored.
- **Wavefunctions are not reality**—they are tools that approximate something deeper.

Physics has fallen into the illusion of quantum weirdness because it stopped asking the right questions.

This book will ask them again.

Because the moment we reject the illusion of probability as fundamental truth, we can finally start looking for the real foundation of the universe.

1.6 The Standard Model: Physics' Biggest Patchwork

For most people, the Standard Model of Particle Physics is celebrated as one of the greatest achievements of modern science. It is presented as the ultimate theory—the fundamental blueprint of reality.

But the truth is much less glamorous.

The Standard Model is not a discovery of nature's fundamental building blocks. It is a catalog of fixes. It is a Frankenstein's monster of equations and particles, stitched together not because reality demanded it—but because our broken theories needed it.

Instead of revealing the truth, it has buried it under layers of mathematical necessity.

The Particle Zoo – When Physics Became a List of Names

Once upon a time, physicists thought they had a simple picture of matter:

- Everything was made of atoms.
- Atoms consisted of protons, neutrons, and electrons.
- Protons and neutrons were solid, indivisible objects.

This was elegant. This was intuitive.

Then, the cracks appeared.

- When high-energy experiments smashed protons apart, new particles appeared.

- Some decayed instantly. Some didn't. Some had no clear purpose.
- More and more experiments revealed more and more unexplained results.

Instead of questioning whether their entire framework was wrong, physicists did what they always do:

- They added fixes.
- They created new particles.
- They built a more complicated model to explain the old one.

And so, the Standard Model was born—not as a discovery, but as a response to failure.

The result?

A zoo of imaginary particles—quarks, gluons, bosons, mesons, and leptons. Each one invented, not discovered, because the equations demanded their existence.

And the worst part?

- None of these particles exist as physical objects in the way protons or electrons do.
- They are mathematical placeholders, forced into reality because without them, the equations collapse.

Instead of admitting, "Maybe we're looking at matter the wrong way," physicists created a fictional subatomic world to fit their models.

The Higgs Boson: A Particle That Had to Exist—Because the Math Said So

In 2012, the discovery of the Higgs boson at CERN was hailed as the final piece of the Standard Model. It was called the "God Particle"—the thing that gave mass to everything.

But here's the truth:

- The Higgs boson wasn't discovered the way electrons or protons were.
- It was mathematically necessary—meaning, if it didn't exist, the Standard Model would fall apart.
- Instead of finding a new law of physics, CERN found exactly what the equations were built to predict.

This is not how science should work.

Imagine a doctor who says:

"Your symptoms don't match any known disease. So I'm going to create a new disease that explains them—and then I'll prove I'm right by diagnosing you with it."

That is exactly what happened with the Higgs boson. It had to exist, because if it didn't, everything physicists had built for decades would be meaningless.

And yet, its discovery didn't explain why mass exists—it just confirmed that our equations were internally consistent.

- It didn't tell us what mass is.
- It didn't tell us why matter resists motion.
- It didn't reveal the deep connection between mass and energy.

It was a bandage—a mathematical trick to keep the Standard Model from collapsing under its own contradictions.

The Standard Model Doesn't Predict Reality—It Adjusts to Fit It

The true test of any theory is its ability to predict new discoveries.

The Standard Model doesn't do that. Instead, it is constantly adjusted every time a new experiment shows something unexpected.

- The neutron's decay wasn't behaving correctly? Add another interaction.
- The weak force doesn't match expectations? Introduce W and Z bosons.
- The universe's expansion doesn't fit the model? Invent dark energy.

This is not how real science works.

The Standard Model is not a framework of reality—it is a list of corrections.

And yet, physicists continue to defend it as if it is the final answer.

Why?

Because to abandon it would mean admitting that physics has spent the last 50 years chasing ghosts.

The Missing Pieces That No One Wants to Talk About

Despite all the patches, the Standard Model still doesn't answer the most fundamental questions.

- What is mass, really? (The Higgs boson is just a mechanism, not an explanation.)
- Why does the universe have matter instead of antimatter?
- What is dark matter, and why can't we detect it?
- Why doesn't the Standard Model include gravity?

The Standard Model ignores these questions because it was never designed to answer them.

It was designed to make the equations work.

The Standard Model Is a Dead End

Instead of being a step toward deeper understanding, the Standard Model has become a prison.

- It forces physics to keep adding new particles and interactions instead of rethinking the foundation.
- It keeps scientists locked into quantum field theory, even though it contradicts relativity.
- It stops physicists from asking the dangerous questions that could expose the truth.

The real problem is not that the Standard Model is incomplete.

The real problem is that physicists refuse to admit it's broken.

- They are too invested in the equations.
- They are too afraid to challenge the foundation.
- They are too comfortable with a patchwork model that "works" but explains nothing.

But science should not be about protecting old ideas.

It should be about questioning everything—especially the models we think are correct.

The Next Section: Breaking Free From the Standard Model

If we want to find the truth, we must break free from the illusion that the Standard Model is the final answer.

- We must stop treating particles as fixed objects and start looking at energy structures.
- We must stop patching equations and start questioning space, mass,

and force at their core.

- We must tear apart the current framework and rebuild it on principles that actually explain reality.

Because as long as we cling to the Standard Model, physics will remain a collection of imaginary fixes—trapped in a world of made-up particles.

And reality deserves better.

1.7 General Relativity: The Mathematically Perfect Fix

Science has a habit of turning convenient models into sacred truths. General Relativity is the perfect example of this.

It is praised as the most elegant theory in physics—a masterpiece of mathematical precision that explains gravity, predicts black holes, and describes the bending of light. But beneath its brilliance lies an uncomfortable truth:

- Einstein knew space-time was just a convenient model.
- Relativity didn't solve gravity—it only redefined it mathematically.
- Even today, physicists still don't know what gravity actually is.

General Relativity didn't fix our understanding of gravity. It buried it under an equation.

The Problem Einstein Had to Solve

Before Einstein, Newton's laws ruled physics. For over two centuries, gravity was understood as a force between masses, described by Newton's equation:

$$F = \frac{Gm_1 m_2}{r^2}$$

It worked beautifully. It explained planets, orbits, and falling apples with remarkable accuracy.

But then, cracks appeared.

- Mercury's orbit didn't match Newtonian predictions.
- Light didn't behave as expected near massive objects.
- There was no mechanism explaining how gravity acts at a distance.

Newton's model was a brilliant description of gravity, but it never explained what gravity actually is.

This was the problem Einstein inherited.

Instead of trying to fix Newton's gravity, Einstein did something radical—he replaced it.

The Space-Time Trick: How Einstein Solved Gravity Without Explaining It

Einstein realized that if he curved space-time itself, he could describe gravity without needing an invisible force acting at a distance.

- Instead of saying, "Mass attracts mass," he said, "Mass bends space, and objects follow the curves."
- Instead of asking why gravity works, he created a model where gravity was just geometry.
- Instead of searching for a physical mechanism, he designed an equation that perfectly matched observations.

It was a brilliant trick.

- No more "mystical" force acting across space.
- No need to explain what carries gravity.
- Everything fit beautifully into a single equation.

Relativity was the mathematically perfect fix. And because it worked, no one questioned it.

The Great Deception – Why Relativity Became Untouchable

Relativity didn't just fix gravity—it killed all competing ideas.

Newtonian gravity had been struggling with contradictions. Scientists had tried for years to tweak it, but nothing worked. Then Einstein came along, and his theory did more than just fix Newton—it made it obsolete.

- General Relativity predicted Mercury's orbit perfectly.
- It explained why light bends near massive objects.
- It even predicted gravitational waves, which were confirmed a century later.

It was so successful that it became unquestionable.

But there was a dangerous side effect—physicists stopped looking for what gravity actually is.

Instead of asking:

- "What is space-time, really?"
- "What is mass doing to space?"
- "Is gravity a fundamental force or just an effect of deeper physics?"

They accepted the mathematical structure as if it were the final truth.

"The math works, so gravity must be curved space-time. Case closed."

But is it?

The Unanswered Mystery: What is Gravity?

Even today, nobody knows what gravity actually is.

- Is it a fundamental force? We don't know.
- Does it have a particle (graviton)? We've never found one.
- Does it fit with quantum mechanics? No, it completely contradicts it.

If General Relativity were truly the final answer, it should connect seamlessly with quantum mechanics. Instead, it breaks down at the smallest scales.

- Inside black holes, Einstein's equations collapse into singularities.
- At quantum levels, gravity doesn't behave like the other forces.
- Despite a century of research, no one has unified gravity with quantum physics.

This is the great failure of relativity:

- It explains gravity mathematically but not physically.
- It describes space-time curvature but doesn't explain what space-time actually is.
- It fits the observations but leaves the mechanism behind gravity a complete mystery.

Why We Need to Look Beyond General Relativity

Einstein himself suspected that General Relativity wasn't the final answer.

- He spent the last 30 years of his life searching for a deeper theory—one that could unify gravity with electromagnetism.
- He never believed in singularities (which his own equations predicted).
- He suspected that space itself must have an underlying structure.

But after his death, physics took a different path.

Instead of questioning relativity's gaps and contradictions, physicists did what they always do:

- They doubled down on the math.
- They invented dark matter and dark energy to fix unexplained gaps.
- They kept relativity as sacred—even when it clearly didn't fit quantum physics.

And so, the mystery of gravity remains unsolved.

The Next Section: Breaking Free from Space-Time

If we want to understand gravity, we must stop treating General Relativity as untouchable.

- We must stop worshiping equations and start questioning reality.
- We must look deeper into the nature of space itself.
- We must rethink gravity—not as a geometric effect, but as an energy interaction.

Because as long as we cling to relativity, we will remain trapped in a mathematical illusion—believing we understand gravity when, in reality, we still have no idea what it actually is.

1.8 The Black Hole Mystery: Hiding Physics' Greatest Failures

For decades, black holes have been treated as the ultimate proof of Einstein's General Relativity—a triumph of theoretical physics. They are portrayed as cosmic monsters, swallowing everything in their path, distorting space-time, and bending reality itself.

But beneath this grandeur lies an uncomfortable truth:

- Black holes are the biggest "we don't know" answer in physics.
- Their very existence exposes contradictions between quantum mechanics and relativity.
- They are filled with paradoxes that science cannot explain—so instead of solving them, physicists simply ignore them.

Rather than being a symbol of physics' success, black holes are a monument to its failures.

The Ultimate Fix – When Physics Created an Object to Hide Its Mistakes

Black holes began as a theoretical accident—a strange prediction that came from Einstein's equations.

- According to General Relativity, if a massive enough star collapses, its gravity becomes so strong that nothing—not even light—can escape.
- As more mass falls inward, the density approaches infinity.
- Space-time curves so extremely that it forms a "singularity"—a point of infinite density and zero volume.

At first, even Einstein didn't believe in black holes. The idea of an infinitely dense, infinitely small object seemed absurd. But physicists, instead of questioning the math, embraced it.

- They accepted singularities, even though physics forbids infinities in nature.
- They accepted event horizons, even though nothing else in physics behaves this way.
- They accepted black holes as real objects, even though nobody actually knows what's inside them.

Instead of admitting, "This makes no sense," physicists declared:

"Black holes exist—because the math says so."

And just like that, an incomprehensible mystery became a scientific truth.

The Paradox of Infinite Density and Zero Volume

Physics has one fundamental rule: No physical quantity can be infinite.

Yet black holes are built on infinities.

- The singularity at a black hole's center is said to have zero volume but infinite density—a mathematical impossibility.
- The event horizon acts as a one-way door—nothing can escape, but how does space-time "know" where to stop?
- General Relativity predicts that time stops completely at the singularity—but what does that even mean?

Whenever infinities appear in physics, it usually means the theory is incomplete or broken.

- When early quantum mechanics predicted infinite energy in atoms, we fixed it with quantum electrodynamics.
- When classical physics failed to explain black body radiation, we fixed it with Planck's quantization of energy.

But when black holes predicted an impossible singularity, we didn't fix it—we just ignored the problem.

Physicists treat singularities not as a failure of relativity, but as something "beyond physics."

"The laws of physics break down inside a black hole."

This is just another way of saying, "We have no idea what's going on."

Yet black holes remain central to modern astrophysics, despite the fact that they expose the biggest failure of our understanding of the universe.

Black Holes vs. Quantum Mechanics – The Deepest Contradiction in Physics

If General Relativity were truly correct, it should seamlessly integrate with quantum mechanics.

But when you try to apply quantum theory to black holes, everything falls apart.

- **Hawking Radiation** – Quantum physics predicts that black holes should evaporate, slowly losing mass over time. But relativity predicts no such thing.
- **The Information Paradox** – Quantum mechanics states that information can never be destroyed. But black holes, according to relativity, erase all information that falls inside them.
- **Gravity vs. Quantum Fields** – Inside a black hole, the gravitational field is so extreme that quantum mechanics stops making sense.

Physicists pretend these contradictions don't exist because they don't know how to fix them.

- Instead of questioning whether space-time is even real, they force relativity and quantum mechanics to coexist—even though they don't.
- Instead of exploring whether gravity is something deeper than curved space, they add more mathematical patches.
- Instead of admitting black holes expose physics' greatest weakness, they celebrate them as proof of Einstein's genius.

The truth is, black holes are not a discovery—they are a confession of our ignorance.

What Black Holes Really Tell Us About the Failure of Modern Physics

The reason black holes remain mysterious is because physics refuses to question its own foundation.

- General Relativity and Quantum Mechanics contradict each other, yet both are treated as absolute truths.
- The singularity is an obvious sign that our model of gravity is incomplete, yet it remains untouched.
- Black holes demand a new theory—one that unifies gravity, quantum physics, and energy.

Yet instead of searching for that deeper truth, physicists continue to defend a model that is mathematically elegant but physically meaningless.

Black holes are not just cosmic mysteries—they are a reflection of how modern physics avoids solving its biggest problems.

They tell us that something fundamental is missing from our understanding of space, mass, and energy.

They prove that our current theories are, at best, incomplete, and at worst, fundamentally flawed.

And as long as we refuse to fix physics at its core, black holes will remain unsolved—not because they are mysterious, but because we are unwilling to see what they truly reveal.

The Next Section: The Dark Matter and Dark Energy Patches

If black holes are the first great confession of physics' failure, then dark matter and dark energy are the second and third.

- They were invented to explain why the universe doesn't behave as expected.
- They have never been detected, yet they dominate physics.
- They are proof that physics continues to adjust equations rather than question assumptions.

In the next section, we will expose the truth behind dark matter and dark energy—and why they are nothing more than modern versions of the ether.

1.9 Dark Matter, Dark Energy: The Last-Resort Fixes

For centuries, whenever physics encountered a problem it couldn't explain, scientists invented something invisible to fill the gap.

- In the 19th century, ether was proposed as an invisible substance filling all space, carrying light waves—because physicists couldn't accept that light could travel through "nothing."
- In the 20th century, virtual particles were introduced—ghost-like entities that pop in and out of existence to fix broken equations in quantum mechanics.
- And now, in the 21st century, we have dark matter and dark energy—the greatest invisible substances ever invented.

Instead of admitting "we don't know how gravity works" or "we are missing something fundamental about space", physicists filled the gaps with two undetectable, unexplained entities.

Dark matter and dark energy are not discoveries.

They are modern-day placeholders for ignorance.

The Dark Matter Problem – When Gravity Didn't Behave as Expected

The story of dark matter began in the 1930s when astronomer Fritz Zwicky noticed that galaxies in the Coma Cluster were moving in ways that Newtonian gravity couldn't explain.

- Based on visible matter, these galaxies should have flown apart.
- Their motion suggested that there was much more mass than what we could see.
- Instead of questioning whether gravity works differently at large scales, scientists declared:

"There must be invisible matter we can't detect."

And so, dark matter was born—not because it was discovered, but because physics needed it to make the math work.

Over the decades, physicists found more discrepancies in galactic rotation curves and cosmic structure. Every time gravity didn't behave as expected, instead of questioning the theory of gravity itself, they added more dark matter.

- Dark matter makes up 85% of the total mass of the universe.
- Yet, after decades of searching, not a single dark matter particle has been found.
- Every experiment designed to detect dark matter has failed.

And still, physicists refuse to consider that gravity itself might work differently than Einstein's equations suggest.

Dark matter is not science—it is a mathematical necessity forced into existence to prevent rethinking gravity.

Dark Energy – The Fix for an Expanding Universe That Shouldn't Exist

If dark matter wasn't bad enough, physics needed another invisible entity—one even more mysterious.

- In 1998, astronomers discovered that the universe's expansion was accelerating.
- This made no sense—according to General Relativity, gravity should slow expansion down.
- Instead of questioning whether Einstein's equations were incomplete, physicists invented dark energy—a mysterious force that pushes the universe apart.

Dark energy now makes up 70% of the universe. It has never been observed or measured. No one knows what it is or why it exists.

And yet, it is accepted as a fact.

Why?

Because without it, relativity fails to explain the universe.

Why Dark Matter and Dark Energy Are the New Ether

The real problem is not that dark matter and dark energy haven't been detected.

The problem is why they were invented in the first place.

- Instead of questioning our understanding of gravity, we added invisible substances to fix the equations.
- Instead of admitting we might be missing something fundamental about space, we forced new concepts into physics.
- Instead of rethinking relativity, we invented magic forces to keep it alive.

This is exactly what happened with ether in the 19th century.

- Physicists believed light needed a medium to travel through, so they invented ether.

- When experiments failed to detect ether, instead of abandoning the idea, scientists kept modifying it.
- It wasn't until Einstein came along that ether was finally discarded.

And now, we are making the same mistake with dark matter and dark energy.

The only difference?

- Ether lasted 50 years before being abandoned.
- Dark matter and dark energy have lasted nearly a century—and still have no proof of existence.

If physics were honest, it would admit:

- We don't understand gravity.
- We don't understand space.
- Dark matter and dark energy are just modern-day ether.

But instead of fixing the foundations, physicists keep adding more mysteries.

The Real Problem: Physics Keeps Patching Instead of Solving

Dark matter and dark energy exist for one reason only:

- To protect General Relativity from being questioned.
- To avoid admitting that Einstein's equations might be incomplete.
- To maintain the illusion that physics has everything under control.

Instead of asking:

- Is gravity actually a force, or is it an effect of deeper energy interactions?
- Does space itself have structure that we've ignored?
- Are we wrong about the fundamental nature of mass and energy?

Physicists avoid these questions because answering them would mean rebuilding physics from the ground up.

So, they cling to dark matter and dark energy—two of the greatest scientific placeholders in history.

The Next Section: The Silent Collapse of Physics

Dark matter and dark energy expose a bigger problem:

- Physics is no longer about discovery—it's about protecting old models.
- The most fundamental questions are ignored to keep equations working.
- The very structure of science has changed—from curiosity to dogma.

In the next section, we will uncover the silent collapse of modern physics—and why the fear of being wrong has kept science stuck for over a century.

Because until we stop adding fixes and start questioning the foundation, physics will remain an illusion—a house of equations built on nothing.

1.10 The Silent Collapse of Physics

For centuries, physics was driven by curiosity and the fearless pursuit of truth. Every great scientist—from Galileo to Newton, Maxwell to Einstein—challenged the old ways, questioned accepted knowledge, and reshaped human understanding.

But today, physics is not about discovery—it is about defense.

- Physicists know that the Standard Model is incomplete.
- They know that General Relativity contradicts quantum mechanics.
- They know that dark matter and dark energy are placeholders, not

explanations.

Yet, they refuse to say it out loud.

Modern physics is experiencing a silent collapse.

Instead of solving fundamental problems, physicists spend their careers protecting old models, adding more patches, and avoiding the real questions.

The search for truth has been replaced with the maintenance of broken theories.

Why Physicists Know That Physics Is Broken (But Won't Admit It)

Every physicist understands that real progress requires questioning the foundation.

Yet, in today's scientific world, anyone who dares to challenge the mainstream narrative is dismissed.

- Ask if the Standard Model is wrong? You won't get funding.
- Suggest that gravity might not be a space-time curvature effect? You'll be ignored.
- Propose that mass and energy are structured in a way we haven't considered? You'll be labeled a crackpot.

Instead of encouraging radical thinking, the physics community has built walls to protect existing theories.

Why?

- Because admitting that physics is broken would mean accepting that an entire century of work needs to be re-evaluated.
- Because questioning Einstein, Bohr, or Dirac is career suicide in a field where tradition is worshipped.

- Because billions of dollars have been invested in finding things like dark matter, supersymmetry, and extra dimensions—and admitting they don't exist would be an embarrassment to the scientific establishment.

And so, physics continues as if everything is fine.

- New particles are predicted, but none are found.
- New experiments are designed, but they only confirm existing models.
- New equations are written, but they never explain anything new.

It is not a search for truth—it is an exercise in protecting the past.

How Research Papers Have Become Exercises in Defending Old Models

If you read a modern physics paper, you will notice a pattern:

- Almost every study starts with the assumption that existing models are correct.
- If data contradicts the theory, the experiment is reinterpreted to fit the theory.
- If a new anomaly is found, it is labeled as "new physics"—but never seen as a reason to question the old physics.

Physics is supposed to be about discovery.

But today, research papers are not about finding the truth—they are about ensuring the Standard Model survives.

- If dark matter isn't found, physicists propose a new type of dark matter instead of questioning gravity.
- If particle accelerators fail to discover supersymmetry, they just increase the energy levels and try again.

- If contradictions arise in quantum mechanics, they are labeled "quantum weirdness" instead of being solved.

The goal is no longer to understand reality.

The goal is to avoid breaking the illusion that we already do.

The Unspoken Truth: Physics Is No Longer a Unified Science

A unified science means all theories fit together without contradictions.

But physics today is a collection of disconnected fixes.

- Quantum mechanics explains the small, but it makes no sense in the large.
- General relativity explains the large, but it breaks down in the small.
- The Standard Model explains particles, but it ignores gravity.
- Cosmology explains the universe, but it relies on invisible substances like dark energy.

Instead of asking:

- Why do we need separate theories for different scales?
- Why does quantum mechanics not apply to gravity?
- Why do we keep inventing invisible things instead of questioning the equations?

Physics keeps pretending that all these contradictions will somehow be resolved in the future.

But deep down, every physicist knows:

- The deeper we dig, the more contradictions we find.
- The more we adjust the equations, the less sense the universe makes.
- The more experiments fail, the more desperate physics becomes.

Modern physics is not thriving—it is stuck.

And no one wants to admit it.

The Real Problem: Science Has Become Afraid of Being Wrong

In the past, physics was not afraid to be wrong.

- When Newton's gravity failed, Einstein replaced it.
- When classical mechanics broke, quantum physics emerged.
- When Maxwell's electromagnetism couldn't explain atomic structure, new physics was embraced.

But today, physics fears change.

- Theories have become too big to fail.
- Careers are built on defending existing ideas.
- Funding is given to safe, predictable research—not groundbreaking discoveries.

Instead of embracing new ideas, physics attacks them.

Instead of asking "What if we're wrong?", physicists ask "How can we adjust the equations to keep the theory alive?"

This is the silent collapse of modern physics.

A field that was once about truth has become about survival.

What Happens Next? The Need for a Scientific Revolution

If physics is to be saved, it must do what it fears most—admit that the foundation is broken.

- We must stop adjusting old models and start rethinking everything from the ground up.
- We must reject dark matter, dark energy, and other placeholders—and search for what is really happening.
- We must question the sacred beliefs of modern physics—even if it means rewriting the last hundred years of science.

Because as long as we refuse to do this, physics will remain trapped—chasing particles that don't exist, forces that don't make sense, and equations that only work in a fictional universe.

Real progress comes not from fixing old mistakes but from discarding them and finding the deeper truth beneath.

The time has come to fix physics—not by patching it, but by rebuilding it.

This book is the first step.

Key Takeaways

- Modern physics has evolved through a **series of temporary fixes** rather than fundamental discoveries.
- Every time physics encountered contradictions, instead of questioning foundational principles, it **introduced new particles or forces** to "fix" the equations.
- **Space-time curvature** and **quantum mechanics** contradict each other, exposing the limitations of the current framework.
- The Standard Model of physics has grown into a **complex and unnatural patchwork** of particles and forces that fails to explain gravity.
- Physics moved away from seeking **why nature behaves the way it does**, focusing instead on **how to make the math work**, leading to a loss of deeper understanding.
- The belief in mathematical consistency has overshadowed the need for **physical realism**, turning physics into an abstract exercise rather than a search for truth.

What If...? (Speculative Exploration)

- What if **physics had questioned its foundations** instead of adding "fixes" every time an equation failed?
- What if space-time curvature was **never the real cause of gravity**, and instead, gravity emerged from an underlying energy structure?
- What if the **entire Standard Model is just a temporary solution**, and we are due for a complete overhaul of physics?
- What if instead of searching for **more particles**, we searched for a missing principle that governs the universe in a fundamentally different way?

Why Has Physics Relied on Fixes Instead of True Discoveries?

- The success of relativity and quantum mechanics led physicists to **prioritize mathematical accuracy over conceptual clarity**.
- Every time an experiment revealed a contradiction, physics found ways to **adjust equations** rather than challenge its assumptions.
- The **rise of mathematical physics** shifted the focus from physical reality to equation-based models, leading to a loss of **intuitive understanding.**
- Funding and academic inertia kept physics trapped in **protecting old theories** rather than allowing space for disruptive new ideas.

Challenges & Open Questions

- If space-time curvature is just a mathematical model, **what is the true nature of space?**
- If gravity is not a force but an emergent property, **how does mass influence the surrounding energy structure?**
- How can we test the idea that space is a **structured energy field** rather than an empty void?
- If modern physics is fundamentally flawed, **what would a new framework look like?**

Suggested Readings & References

- **Tesla's Theories on Energy and Space** – An alternative to Einstein's space-time model.
- **Einstein's General Relativity and Its Incompleteness** – Why relativity remains an incomplete theory.
- **Chavan's Grand Unified Theory of Physics** – A new approach that challenges traditional physics.

2: The Space-Time Fantasy

Einstein's Necessary Fix

Physics had reached a breaking point.

By the early 20th century, Newtonian gravity—once considered absolute truth—was failing. It could not explain Mercury's orbit, nor could it answer the deeper question that had haunted physicists for centuries: How does gravity actually work?

Newton described gravity as a force acting at a distance, but he never explained what carried that force. How could the Sun pull on the Earth without any physical connection between them? Was there an invisible medium filling space that transmitted this pull, or was gravity something else entirely?

Physicists once believed in ether, a structured background energy that filled space and allowed forces to act across distances. But when experiments failed to detect it, science took a radical step—it declared space to be truly empty.

This left Einstein in a difficult position. He had already revolutionized physics with Special Relativity, proving that space and time were not separate but connected as a four-dimensional entity. Now, he needed a new explanation for gravity—one that didn't require a hidden medium.

Instead of uncovering the physical cause of gravity, Einstein redefined space itself. He made gravity not a force, but a geometric effect. He replaced the idea of an energy-filled space with a mathematical construct called space-time. He curved space instead of explaining what mass actually does to space.

It was brilliant. It was elegant. It worked mathematically.

But was it real?

Or was it simply a necessary fix to keep physics moving forward?

This chapter exposes the hidden assumptions behind Einstein's space-time—how it was born, why it was accepted, and why gravity is not curvature, but an energy effect that physics has ignored for over a century. The story of space-time is not just about Einstein's genius—it is about how physics took a convenient detour instead of searching for the true nature of gravity.

Now, let's go back to the moment Einstein faced the greatest challenge of his career.

2.1 Einstein's Dilemma: The Missing Medium for Gravity

By the early 20th century, physics stood at a crossroads. Newton's laws had governed the cosmos for centuries, allowing humanity to predict planetary orbits, the motion of falling objects, and the tides that danced to the Moon's silent pull. But beneath its precision lay an unsettling void. Newton had described gravity—but he had never explained it.

How does the Sun pull on the Earth across empty space?

For all its mathematical beauty, Newtonian gravity lacked a physical mechanism. How could one celestial body influence another millions of kilometers away without an invisible tether connecting them? This question had lingered in the background for generations. Some scientists believed that space itself had a hidden structure—a medium that carried gravitational interactions, just as air carried sound or water carried waves. They called it ether.

But in 1887, the famous Michelson-Morley experiment failed to detect this mysterious substance. The results sent shockwaves through physics. If ether did not exist, then space must be truly empty, a void where light traveled without a carrier and forces acted with no intermediary. The scientific community rushed to embrace this conclusion, discarding the very notion that space could have structure.

Then came Einstein.

In 1905, his theory of Special Relativity shattered long-standing assumptions about motion and time, proving that nothing—not even information—could travel faster than light. But this introduced an even greater problem: Newton's gravity, in its original form, required instantaneous action at a distance. If the Sun were to suddenly vanish, Newton's equations predicted that the Earth would immediately fly off into space, as if the Sun had never existed. That would mean gravity traveled infinitely fast—a direct violation of Einstein's new framework.

Something had to change.

If gravity was not an invisible force acting instantaneously across space, then what was it? What carried its influence? The answer Einstein proposed would redefine our understanding of reality itself.

Instead of treating space as an empty stage upon which physics played out, Einstein made it an actor in the drama. He redefined space and time as a single, flexible entity: space-time. It was no longer a passive void but a dynamic fabric that could be stretched, compressed, and curved by the presence of mass and energy. In doing so, he found his answer: gravity was not a force at all, but an illusion created by the warping of this four-dimensional continuum.

Where Newton had described a universe governed by forces, Einstein replaced them with geometry. The Sun did not pull on the Earth. Instead, it bent the space-time around it, and the Earth simply followed the contours of this invisible curvature. It was an elegant solution, one that explained planetary motion, gravitational lensing, and even the anomalies in Mercury's orbit that Newton's laws had failed to predict.

It was beautiful. It was revolutionary.

But was it real?

Or was it simply the most sophisticated mathematical fix in the history of physics?

2.2 The Birth of Space-Time: A Convenient Fiction

The greatest trick modern physics ever pulled was convincing the world that space-time is real.

Before Einstein, space and time were considered separate, absolute entities—a fixed backdrop where the universe played out its motions. Newton's space was an unchanging stage, independent of forces and movement, while time flowed uniformly for all.

Then Einstein's equations arrived, shaking the foundations of classical physics. He needed to explain why gravity didn't behave like other forces, why massive objects influenced motion without a physical medium, and why Mercury's orbit refused to follow Newton's predictions.

His solution was not to uncover a physical cause for gravity but to redefine space itself. He turned time into a dimension and replaced force with geometry. Instead of explaining what mass actually does to space, he bent space and time itself.

This wasn't a discovery; it was an invention.

And it worked. The equations fit beautifully, predictions aligned with observations, and soon, the world stopped questioning whether space-time was real and simply accepted it as truth.

When Time Became a Dimension – Not Because It Is, But Because It Made the Math Work

Einstein's most radical shift was not just merging space and time, but elevating time to the status of a fourth dimension.

In Newtonian mechanics, time was independent of space. In Special Relativity, time became flexible—it could slow down or speed up depending on motion. In General Relativity, time was fully merged into space, forming a four-dimensional continuum called space-time.

But was this really how the universe works?

We experience space as something physical—we can move through it, measure it, and interact with it. But we experience time as an emergent effect—we cannot move backward, we cannot hold it in our hands. Unlike space, time has no physical existence—it is merely a measurement of change.

Yet, in Einstein's equations, time was treated as just another coordinate.

Why? Because it made the equations work.

Space-time was not discovered through observation—it was formulated out of necessity. If Einstein had treated time as an independent quantity, his model of gravity would have collapsed. To make gravity a geometric effect, he had to treat time as a spatial dimension—even though it isn't.

Thus, time was merged into space, not because reality demanded it, but because mathematics required it. A brilliant illusion, but an illusion nonetheless.

Space-Time Curvature: A Theory Built to Save Relativity

If gravity wasn't a force, Einstein needed a new explanation for why masses move the way they do. Newton said the Sun pulls on the Earth. Einstein said the Sun bends space-time, and the Earth follows the curve.

It was an elegant solution. But here's the key problem: space cannot bend—because space is not a physical substance. Time cannot bend—because time is not an object.

Yet relativity demands that we believe this is exactly what happens.

When an object moves, it "follows the curvature" of space-time. When a planet orbits a star, it isn't being pulled—it is simply "rolling along a distorted path." When light bends around a massive object, it isn't being affected by gravity—it is following the shape of space-time.

This is not physics—it is an elegant mathematical construct.

Instead of asking what mass actually does to space, whether gravity has an energy-based cause, or if there could be a deeper structure to space itself, physics took the easy way out and declared:

Mass bends space, and space tells objects how to move.

This was not a discovery—it was a patch to save relativity from its own contradictions.

Mathematical Elegance vs. Physical Reality

Space-time is beautiful on paper. The equations are simple, powerful, and precise. They can predict planetary motion, gravitational lensing, and even black holes.

But here's the problem: space-time is not a physical entity—it is a coordinate system. It cannot curve, warp, or bend—because it has no physical properties. Mass cannot bend space—because space is not an object that can be bent.

Imagine drawing a road map. You might mark curves, elevations, and distances. But does the map itself shape reality? If you draw a curve on a piece of paper, does the road actually curve in reality? If you stretch the graph of a function, does the real world stretch along with it?

Space-time is a mathematical map, not a tangible structure.

Einstein's equations describe how masses move, but they do not tell us why mass interacts with space at all. Yet physics continues to treat space-time as if it were a real, physical object, instead of asking the deeper question: If gravity is not bending space, what is actually happening?

The Mistake That Changed Physics for a Century

When Einstein introduced space-time, it solved key problems in Newtonian gravity. It explained Mercury's orbit. It predicted gravitational lensing. It provided a new mathematical framework for physics.

But in doing so, it created even bigger problems—problems that physics still refuses to face. Gravity still has no real cause. Mass and space have no clear relationship beyond equations. Black holes, singularities, and infinite densities all expose contradictions in space-time.

Instead of treating space as an active medium with energy density, Einstein made it a coordinate system that could bend. Instead of searching for a deeper energy-based mechanism for gravity, he curved space-time to make the equations work.

And now, physics is stuck. Quantum mechanics and relativity do not agree. Gravity is still not unified with the other forces. The structure of space remains completely unknown.

Space-time is a beautiful illusion, but an illusion nonetheless.

If physics wants to move forward, it must reject the fantasy of curved space and start searching for the real mechanism behind gravity. Because as long as we believe in space-time, we will remain trapped in a mathematical dream—mistaking equations for reality and ignoring the deeper truth hidden beneath them.

Next: The Hidden Assumption – Space is Not Empty, It is Structured Energy

If space-time is just a convenient fiction, then what is the real structure of space? Is space truly empty, or does it have an energy density that Einstein ignored? Could gravity be an energy effect, rather than a geometric curvature? Has modern physics unknowingly reintroduced a structured vacuum under new names—like quantum fields, dark energy, and the Higgs field?

In the next section, we expose the greatest mistake in physics—the assumption that space is empty, when in reality, it is a structured, energy-filled medium that may hold the true secret of gravity.

2.3 The Hidden Assumption: Space is Not Empty, It is Structured Energy

One of the greatest misconceptions in modern physics is the belief that space is empty.

For centuries, scientists assumed that space must contain something—a medium that allowed forces to act across vast distances. This was the natural way of thinking: sound needs air, waves in water need a liquid surface, and electric and magnetic forces act through fields.

Yet, when it came to gravity, physics made an exception. Newton described gravity as an invisible force acting across space, but he never explained what carried that force. Einstein discarded the idea of a physical medium, redefining space itself as a mathematical construct. Instead of asking what space actually is, physics simply declared it a vacuum—a void with no physical properties.

This assumption was not grounded in evidence—it was a necessity to make relativity work. And because of it, physics turned away from the real answer. Space is not nothing—it is an energy-filled medium. Gravity is not a bending of coordinates—it is an interaction with this structured energy. Einstein did not uncover gravity's mechanism—he obscured it with space-time curvature.

Modern physics has quietly reintroduced vacuum energy under different names, disguising the fact that space was never truly empty.

The Greatest Mistake in Modern Physics: Declaring Space is Empty

If there is one fundamental flaw that has confined physics for a century, it is the belief that space is truly a vacuum. Newton treated space as a passive background, but left gravity unexplained. Einstein abandoned the ether but replaced it with space-time—a purely mathematical construct. Quantum mechanics later discovered that "empty space" is filled with fluctuations and energy, yet relativity still treats it as a void.

This contradiction is largely ignored in physics. On one hand, relativity treats space as a smooth, continuous fabric that can stretch and bend—yet this same space is considered a vacuum in every other area of physics.

But if space is truly empty, then why does quantum mechanics say space is filled with virtual particles? Why does dark energy exist if space has no intrinsic properties? Why does gravity behave like a field interaction rather than a geometric effect?

The answer is simple: space is not empty. It never was.

Modern physics has been forced to bring back the concept of vacuum energy—but instead of admitting that space has structure, it has been disguised under new terminology.

The Quiet Return of Vacuum Energy – Disguised as Quantum Fields and Dark Energy

When Einstein rejected the ether, he unknowingly dismissed the possibility that space itself has an intrinsic energy density. But as physics advanced, this assumption became impossible to maintain. Quantum mechanics revealed that empty space is filled with energy fluctuations. The Higgs field was introduced—an invisible field that fills all space and gives particles mass. Dark energy was proposed as a mysterious force that fills space and accelerates the universe's expansion.

One by one, physics was forced to reintroduce structured vacuum energy—without calling it that.

The result? Quantum fields now exist everywhere, but we still call space "empty." Dark energy is said to make up 70% of the universe, but we still claim vacuum has no structure. Gravitational waves propagate through space like waves in a medium, yet we still insist space has no physical substance.

This is the silent contradiction at the heart of modern physics. Instead of acknowledging that space is an active energy field, physics has invented patch after patch to avoid admitting it.

Every Force in Physics Works Through a Medium—Except Gravity?

A fundamental principle of physics is that forces require a medium to act through. Electromagnetic waves move through the electromagnetic field. The strong nuclear force is carried by gluons in a structured energy field. The weak nuclear force is mediated by W and Z bosons.

Yet, gravity remains an exception. Relativity claims that gravity is not a force—it is space-time curvature. But quantum mechanics treats forces as energy exchanges between fields. If gravity is a real effect, why doesn't it have a carrier or a medium?

This is the greatest inconsistency in physics. If all forces work through fields and interactions, gravity should not be different. The idea that gravity is merely the bending of coordinates is an assumption that has never been tested. If space is structured energy, then gravity must be an interaction with that energy—not a distortion of geometry.

This is the key insight physics has overlooked for a century. Gravity is not a mysterious, unique bending of space-time. It is an interaction with a real, physical structure of space—an energy field that exists throughout the universe. It behaves like a field because it is one. It does not require imaginary curvature because it follows energy gradients.

Once we reject the illusion of space-time, we can finally begin to understand what gravity actually is.

Rebuilding Gravity from the Ground Up

If physics wants to move forward, it must abandon the idea that space is nothing. Space is an energy-filled medium. Gravity is an interaction with this structured vacuum energy. Relativity's space-time curvature is a mathematical illusion, not a physical reality.

This means the real solution to gravity lies in understanding how mass interacts with this hidden structure of space. Why does mass affect space's energy density? How does space itself store and transfer energy? Is gravity actually an emergent effect of energy gradients, rather than a fundamental force?

The answers to these questions have been ignored for too long. Now, it is time to correct that mistake.

Physics has spent a century believing in space-time. But the truth is far more profound: gravity is an energy interaction, not a curvature of coordinates.

Next: The Curvature Myth – Why Gravity is an Energy Gradient, Not Bent Space

If space is a structured energy field, then gravity must be an interaction with that energy—not a warping of space-time. Why do gravitational waves behave like waves in a medium? Why does mass seem to "bend" space if space is actually an energy structure? How can we redefine gravity as an energy effect rather than a geometric illusion?

In the next section, we will break the final illusion—that space-time is curved—and replace it with the real mechanism behind gravity. Because once we see gravity for what it truly is, physics will never be the same again.

2.4 The Curvature Myth: Why Gravity is an Energy Gradient, Not Bent Space

One of the most successful misconceptions in modern physics is the belief that gravity is the curvature of space-time.

Einstein's equations beautifully describe how mass influences motion, but they do not explain what gravity actually is. Why does mass bend space-time? What is space-time made of that allows it to bend? Why does gravity behave like an energy interaction rather than a geometric effect?

For over a century, physics has ignored these questions, treating relativity as the final answer. But space-time curvature is not a discovery—it is a convenient mathematical illusion. It describes gravity in terms of geometry but does not reveal the physical mechanism behind it. It works mathematically, but it collapses in extreme cases like black holes and singularities. It provides no connection to quantum mechanics, making gravity the only force in physics with no clear underlying structure.

It is time to expose the fundamental flaw of space-time curvature—and reveal why gravity is not a warping of coordinates, but an energy gradient in a structured medium.

The Fundamental Flaw of Space-Time Curvature: It Describes Gravity, But Does Not Explain It

General Relativity tells us that mass bends space-time, objects move along curved paths in this distorted space, and gravity is not a force but the result of this curvature. But ask yourself: what is space-time made of that allows it to bend? If mass distorts space, then what is it actually interacting with? Why does this curvature produce effects identical to an energy gradient?

Space-time curvature is a description, not an explanation. It tells us what happens when mass is present but not why it happens. Newton's gravity was criticized because it described attraction but did not explain the mechanism. Yet, General Relativity does the same thing—it replaces force with geometry but never explains what space-time actually is.

If a model cannot explain its own foundation, it is not the final answer. And the deeper we look, the more gravity behaves like an energy interaction—not a geometric effect.

Gravity Behaves More Like an Energy Field Than a Geometric Distortion

Every other force in physics is understood as an energy-based interaction. Electric fields interact through charges. Magnetic fields interact through moving charges. The strong force interacts through energy exchanges between quarks.

Yet gravity is the only force that is treated differently—not as an energy field, but as a "bending of space."

This assumption has no physical justification. If gravity were purely geometric, there should be no energy exchange in gravitational interactions. But mass-energy equivalence ($E=mc^2$) suggests that mass interacts with space in an energy-driven way. The presence of mass seems to alter the properties of space itself—not just its shape.

This suggests that gravity is not a distortion of empty space—it is an interaction with an underlying energy structure. If space has structure, gravity must be an energy effect within that structure. If mass influences space, it must be affecting something real—not just coordinates. If gravity behaves like a field, it must be treated as one—not as a mathematical abstraction.

General Relativity ignores this, treating space as a mathematical stage rather than a physical entity. But gravitational waves reveal the truth that relativity refuses to acknowledge.

Why Gravitational Waves Prove Space Has Structure, Not Just Curvature

Gravitational waves are ripples in space-time, predicted by Einstein and detected a century later. Physicists celebrate this as proof of relativity—but in reality, it exposes its greatest flaw.

If space-time were just a coordinate system, how could it carry waves? If space were truly empty, what is oscillating when a gravitational wave moves through it? If curvature alone explained gravity, why do gravitational waves behave like energy waves in a medium?

The behavior of gravitational waves suggests that space is not just geometry—it has physical properties. They move through space like waves in a structured medium. They carry energy, which means space must store and transfer energy. They interact with mass, which suggests an underlying structure that mass disturbs.

This is direct evidence that space is not a void—it is an active energy field. If space were truly just a mathematical coordinate system, there would be nothing to wave. The fact that gravitational waves exist proves that space has structure—one that Einstein ignored in order to make relativity work.

Einstein's Equations Break Down in Extreme Cases – The Limits of Space-Time

Relativity is often praised for its accuracy. But what physicists don't talk about is where it fails.

Black holes—Einstein's equations predict a singularity, an infinite curvature of space-time. But infinities in physics always signal a broken theory. The Big Bang—General Relativity collapses at the beginning of time, requiring a patchwork of new theories like inflation and quantum gravity. Quantum Gravity—Relativity and quantum mechanics are incompatible, yet both are treated as fundamental.

Whenever relativity is pushed to its limits, it self-destructs. Singularities make no physical sense. Space-time curvature cannot explain why mass interacts the way it does. Gravity refuses to fit into the quantum framework, exposing the illusion of geometric warping.

Everywhere we look, relativity is telling us it is incomplete. Instead of blindly trusting curved space-time, physics should be asking: What is space actually made of? What if gravity is an energy effect rather than a bending of coordinates? What deeper mechanism is missing from our current models?

Because once we stop believing in the myth of space-time curvature, we can finally start looking for the real answer.

The Next Step: Unmasking the True Nature of Gravity

If space is not just a coordinate system, then gravity must be an interaction within that structure. Mass does not bend space—it affects an underlying energy field. Gravity is not geometry—it is an effect of mass interacting with the vacuum structure of space. Space-time is not the final answer—it is a temporary fix for a missing physics.

The next section will explore what space is actually made of, why gravity follows energy gradients rather than curved paths, and how rejecting space-time leads to a new understanding of mass, inertia, and motion.

The deeper we go, the clearer it becomes: Einstein's greatest achievement was also his greatest compromise. The real physics of gravity has been hidden for a century. Now, it is time to uncover the truth.

2.5 Why the Universe Doesn't Need Space-Time to Explain Gravity

For over a century, physics has clung to space-time curvature as the best explanation for gravity. The idea that mass bends space, guiding objects along curved paths, has been so deeply ingrained in science that few ever question it.

But what if space-time isn't needed at all? What if gravity is not a geometric effect, but an energy gradient? What if mass does not bend space, but instead disturbs an underlying energy structure? What if the missing link between gravity, mass, and quantum mechanics has been hidden in plain sight all along?

Modern physics remains trapped between two incompatible theories—relativity and quantum mechanics. But once we discard the illusion of curved space-time, a true unification of physics becomes possible.

This section exposes why space-time is an unnecessary complication and how gravity can be understood in terms of energy interactions rather than mathematical distortions.

The End of Space-Time: Why Gravity Works Without It

For gravity to be a fundamental force, it should behave like all the other forces in nature. Electricity follows energy gradients—charges move from high to low potential. Magnetism follows field lines—forces are carried through an active medium. Strong and weak nuclear forces follow energy interactions—exchanges between quarks and particles.

Yet, gravity is the only force described as a geometric warping instead of an energy exchange. This inconsistency makes no sense. If mass affects space, it must be interacting with something physical—not just a coordinate system. If gravitational waves carry energy, they must propagate through a real medium—not an abstract fabric. If space is curved, then what is actually being bent?

The simplest answer? Gravity is not a warping of space—it is an energy gradient in a structured field.

Just as heat flows from high temperature to low temperature, gravity emerges from an energy difference in space itself. This eliminates the need for space-time curvature entirely.

The Missing Link Between Gravity, Mass, and Energy Density

One of the greatest mysteries in physics is why mass creates gravity. Why does mass attract other mass? Why does inertia exist? Why do objects resist acceleration?

General Relativity never explains these things—it merely describes how mass behaves in space-time. But if we discard space-time and treat space as a structured energy field, then gravity finally makes sense. Mass is a concentration of energy. It disturbs the energy balance in space, creating a gradient. Objects move toward lower-energy regions, just as charges move in an electric field.

This means gravity is not a fundamental force—it is an effect of mass interacting with structured energy. Objects do not "bend space"—they alter the energy density of the vacuum, creating an attractive potential. Inertia is simply resistance to changes in this energy balance.

Suddenly, gravity, mass, and inertia are all connected—not through geometry, but through energy interactions.

Why Treating Space as a Structured Energy Field Unifies Gravity with Quantum Mechanics

The greatest failure of modern physics is its inability to unify gravity with quantum mechanics. Relativity describes gravity as smooth space-time curvature. Quantum mechanics describes forces as discrete energy exchanges in a field. These two views contradict each other completely.

But if gravity is an energy gradient in a structured field, then it naturally fits within quantum mechanics. Quantum fields exist everywhere, even in a vacuum. Gravity can emerge from fluctuations in this field, rather than bending space. Mass is not "curving" anything—it is disturbing the energy density of space.

This means there is no need for a graviton—the quantum of gravity is just an energy disturbance in the vacuum. There is no conflict between gravity and quantum mechanics—because they both describe energy interactions in space. The universe does not need a separate force of gravity—because gravity is just an effect of structured energy.

For the first time, gravity can be understood not as an exception, but as part of the same framework as all other forces.

The Future of Physics: From Mathematical Illusions to Physical Reality

Physics has spent a century believing in the fantasy of space-time. Einstein replaced ether with a mathematical construct instead of searching for a real medium. Physicists abandoned structured space in favor of abstract coordinates. Quantum mechanics quietly reintroduced vacuum energy, but without acknowledging its role in gravity.

It is time to correct this mistake. Space is not a void—it is an energy-filled structure. Mass does not bend space—it disturbs the energy density of space. Gravity is not a force—it is an emergent effect of energy gradients.

By rejecting space-time, we can finally move physics forward. Unify gravity with quantum mechanics. Eliminate the need for invisible fixes like dark matter and dark energy. Explain the true nature of mass, inertia, and motion.

The biggest mistake in physics was turning space into a mathematical illusion. The greatest breakthrough will come when we restore space as a structured, energy-filled reality. Because once we do, the true nature of gravity will finally be revealed.

Key Takeaways

- Einstein's **space-time model** replaced Newtonian gravity, but it did not **explain** gravity—it only **redefined it as geometry**.
- The concept of **curved space-time** was a mathematical fix that worked in equations but lacked a true physical mechanism.
- **Space-time is treated as a fabric**, yet physics never defines what it is made of—**is it an entity, a field, or just a coordinate system?**
- The idea that mass bends space creates contradictions, as it **does not explain the true cause of gravity**.
- Despite its success, general relativity fails at **quantum scales and inside black holes**, proving it is incomplete.
- If general relativity is **only an approximation**, then physics needs a **new way to understand gravity** beyond space-time curvature.

What If...? (Speculative Exploration)

- What if **space-time is not a fundamental reality** but just a mathematical tool?
- What if **gravity does not curve space**, but instead emerges from the structured energy within space?
- What if black holes do not have singularities, but instead reveal the **limits of space-time curvature** as a valid model?
- What if Einstein himself **had doubts** about space-time but lacked the tools to explore an alternative?

Why Did Physics Accept Space-Time Curvature Without Question?

- The **Michelson-Morley experiment** led to the rejection of the ether, forcing physics to find an alternative explanation for gravity.
- Einstein's equations **worked** in explaining planetary motion and

light bending, making them **too successful to challenge**.
- The mathematical elegance of relativity made physicists **reluctant to explore other models**.
- Space-time curvature provided a way to **fix Newton's problems** without reconsidering the deeper nature of space itself.
- Quantum mechanics was developing at the same time, and **instead of merging gravity with quantum theory**, physics treated them as separate realms.

Challenges & Open Questions

- If gravity is not curved space-time, **what is the true cause of gravitational attraction?**
- If space-time is not fundamental, **what is the real structure of space?**
- How can we detect **structured vacuum energy** as an alternative to curved space?
- Can an **energy-based model of gravity** explain why mass influences motion without needing space-time curvature?

Suggested Readings & References

- **Einstein's Own Doubts on General Relativity** – A historical look at Einstein's struggle with unifying gravity and quantum mechanics.
- **Tesla's Vision of Structured Space** – How Tesla's ideas on energy fields contrast with space-time.
- **Chavan's Grand Unified Theory of Physics** – An alternative approach that explains gravity without space-time curvature. (https://doi.org/10.5281/zenodo.14791053)

3: The Quantum Cover-Up

The Illusion of Wave-Particle Duality

Modern physics is built on contradictions disguised as discoveries.

In the previous chapters, we dismantled the illusion of space-time, revealing how Einstein's curved geometry was a mathematical construct rather than a physical explanation. Now, we turn to another deeply entrenched paradox—wave-particle duality, the cornerstone of quantum mechanics.

At the dawn of the 20th century, physics faced a profound crisis. The classical understanding of light as a wave failed to explain the photoelectric effect. Instead of reconsidering the nature of space itself or acknowledging that light might interact with an underlying energy field, Einstein introduced photons—discrete particles of light.

Not because light is inherently a particle, but because physics needed an immediate theoretical fix. Instead of questioning whether space contained a structured energy medium, physicists forced light into two contradictory identities—sometimes a wave, sometimes a particle. This gave rise to the illusion of wave-particle duality—a concept that persists not because it provides a fundamental understanding, but because physics never sought a deeper explanation.

Just as space-time curvature was introduced to sidestep the true mechanism of gravity, wave-particle duality became a convenient mathematical workaround—a superficial answer to quantum phenomena that avoided addressing the true nature of light and matter.

In this chapter, we expose the quantum cover-up: why Einstein introduced photons instead of recognizing structured vacuum energy, how quantum mechanics describes reality but fails to explain it, and why wave-particle duality remains an unresolved contradiction that physics refuses to challenge.

It is time to move beyond illusion and uncover the real nature of light, energy, and quantum interactions. The universe does not operate on contradictions—it follows principles that physics has long ignored.

Let's begin by revisiting the moment Einstein made his greatest compromise—the birth of the photon.

3.1 Einstein's Dilemma: Why He Created Photons Instead of Recognizing Structured Vacuum Energy

At the dawn of the 20th century, physics stood at the brink of a revolution. Classical mechanics had successfully explained planetary motion, thermodynamics, and Maxwell's equations had unified electricity and magnetism into a single framework.

But a major problem remained—light, one of the most fundamental phenomena in physics, refused to behave as expected.

For centuries, light had been understood as a wave, confirmed by the famous double-slit experiment, which demonstrated clear interference patterns. Even Maxwell's equations described light as an electromagnetic wave propagating through space. Yet, when physicists examined how light interacted with matter in the photoelectric effect, they encountered an anomaly that defied classical wave theory.

Light striking a metal surface ejected electrons, but only if the light had a high enough frequency. Increasing light intensity did not increase electron energy—only frequency mattered. Classical wave theory predicted that energy should depend on wave intensity, not frequency.

Something wasn't adding up.

This was Einstein's dilemma—how could light, well-established as a wave, behave like a particle when interacting with matter?

Rather than question whether space itself played a role in energy transfer, Einstein took a radical step. He proposed that light wasn't just a wave—it was also a particle. This led to the birth of the photon, a quantum "packet" of light energy.

Einstein's explanation worked. It matched experimental results, solved the photoelectric mystery, and eventually won him the Nobel Prize.

But this solution came with hidden consequences. Instead of exploring whether space had an intrinsic energy structure that absorbed and re-emitted energy, physics hastily declared that light must be a particle. Instead of reconsidering the role of an underlying vacuum energy field, physics introduced photons as fundamental particles. Instead of investigating the nature of space, physics treated it as an empty void—burying deeper questions for a century.

Einstein solved one mystery but created another. And modern physics never looked back.

The Real Question Einstein Avoided: Does Light Need a Medium to Propagate?

Einstein's concept of photons did not emerge from direct observation—it was a theoretical construct meant to resolve an experimental contradiction. But he never addressed a more fundamental question.

If light is a wave, what is waving? Does light require a medium to travel through, like all other waves? If space is truly empty, how does light propagate through it?

For centuries, physicists believed space was filled with an invisible medium called the ether. Ether was thought to carry light waves, just as air carries sound waves. Maxwell's equations implied that electromagnetic waves needed a medium to oscillate within. The famous Michelson-Morley experiment failed to detect ether's motion, leading scientists to abandon the concept.

But rejecting ether did not necessarily mean rejecting the idea that space had structure. The experiment only failed to detect motion of the ether—it did not prove that space was truly empty. Space could still possess an intrinsic energy structure, even if it did not behave like a classical fluid. Instead of investigating this possibility, physics assumed space was a vacuum and moved on.

This was Einstein's dilemma. He needed light to behave as a particle to explain the photoelectric effect. He knew light exhibited wave behavior in other experiments. Instead of questioning the nature of space, he split light into two contradictory behaviors—wave and particle.

This convenient mathematical fix ignored the deeper question.

Why Einstein Ignored Vacuum Energy and Instead Created an Ad Hoc Explanation

Had Einstein considered the possibility that space itself contained energy, he might have avoided the need for photons altogether.

But at the time, physics was moving away from structured space. The failure to detect ether made physicists reluctant to propose any kind of medium. Quantum mechanics was still in its infancy, and the quantum nature of space was not yet understood. Even Einstein later introduced the cosmological constant, hinting at vacuum energy—but he abandoned it.

Rather than questioning whether space might actively participate in energy transfer, Einstein introduced photons as fundamental particles.

This decision shaped modern physics, but it also introduced contradictions that still plague quantum mechanics today.

If photons are particles, why do they create interference patterns in the double-slit experiment? If photons exist as discrete quanta, what happens to their wave nature when they "collapse" into particles? If light is sometimes a wave and sometimes a particle, what determines when it switches?

Rather than answering these questions, physics accepted the contradiction. Wave-particle duality became a foundational concept—when it should have been a red flag. And the debate over the true nature of space was silenced.

How Physics Silenced the Debate by Accepting Photons as Fundamental

The photon was never discovered as a physical particle—it was a theoretical construct meant to explain an anomaly. Yet, once introduced, it became unquestioned dogma in physics.

Quantum mechanics built entire models around photons, treating them as real particles. Particle physics accepted light as a stream of energy packets, despite overwhelming wave behavior. The Standard Model enshrined photons as force carriers, ignoring the possibility of structured space.

Instead of asking whether light is a wave propagating through an unseen energy field, whether vacuum energy could explain why light behaves the way it does, or whether the photon is just a mathematical tool rather than a real entity, physics chose the easy answer—and abandoned the search for the real one.

Now, a century later, we are left with a contradiction that no one dares to resolve. Light is sometimes a wave, sometimes a particle—because we refuse to acknowledge the possibility of a structured vacuum. Quantum mechanics still cannot define what a photon actually is—because it was never physically real to begin with. The assumption that space is empty remains unquestioned—despite overwhelming evidence that space is filled with quantum energy.

This is the quantum cover-up—a fundamental misunderstanding that physics refuses to admit.

Instead of recognizing that light moves through a structured energy field, physics built an entire framework around a fictional particle. This is why quantum mechanics describes reality but fails to explain it. This is why wave-particle duality persists—it is a symptom of an incomplete theory. This is why modern physics remains stuck—because it refuses to question its own foundation.

Next: The Quantum Mechanics Paradox – It Describes Reality, But Fails to Explain It

If photons are merely a mathematical fix, then what is really happening in quantum mechanics? Why does the double-slit experiment create interference, even with single particles? Why does quantum mechanics rely on probabilities instead of clear physical explanations? Why does physics refuse to define what electrons and photons actually are?

In the next section, we reveal how quantum theory became a predictive tool rather than a true understanding of reality. Because as long as physics continues to accept contradictions, it will never uncover the truth.

3.2 The Quantum Mechanics Paradox: It Describes Reality, But Fails to Explain It

Modern physics prides itself on the success of quantum mechanics—a framework so precise that it can predict subatomic interactions with near-perfect accuracy.

But there's a problem. Quantum mechanics describes reality, but it does not explain it. It calculates probabilities but refuses to define what is actually happening. It gives us equations, but no physical interpretation that makes sense.

Physicists will tell you that quantum mechanics is the most successful theory ever created. But success in prediction is not the same as understanding.

At its core, quantum mechanics is not a theory of reality—it is a set of mathematical rules that happen to work. It tells us what to expect, but not why it happens. It uses probability to describe events but avoids defining what is physically occurring. It claims electrons and photons are both waves and particles—but never explains how or why.

Rather than searching for a deeper truth, physics has embraced a "shut up and calculate" approach—ignoring contradictions instead of resolving them.

Nowhere is this failure more obvious than in the most famous experiment in quantum mechanics—the double-slit experiment.

The Mystery of the Double-Slit Experiment: Where Does the Wave Come From?

The double-slit experiment is the foundation of quantum mechanics. It revolutionized physics—yet remains one of its greatest mysteries.

When particles (electrons or photons) pass through two slits without measurement, they form an interference pattern—like waves. But when we measure which slit they pass through, the interference pattern disappears—they behave like particles. Somehow, the act of "observation" determines the outcome.

This is the quantum paradox. When unobserved, the electron behaves like a wave. When observed, it behaves like a particle. But where does the wave come from in the first place?

Physics has spent a century dodging this question. The standard response is that "the wave is just a probability function." But that doesn't explain why probability behaves like a physical wave. Nor does it explain what is waving—because if space is empty, there should be nothing to carry the wave.

Instead of answering these questions, quantum mechanics chose the easy way out. It declared, "The wave function is not real—it is just a tool for calculations."

This is an abdication of science. Instead of defining what an electron or photon actually is, physics hides behind mathematics. Instead of asking what is physically happening, it turns probability into a stand-in for reality. Instead of investigating the true nature of space, it ignores the possibility that space itself has quantum properties.

The result? A mathematical model that predicts outcomes—but doesn't explain what is actually occurring. A paradox that remains unsolved—not because it can't be solved, but because physics refuses to rethink its assumptions. A century of quantum mechanics treating contradictions as facts instead of recognizing them as signs of an incomplete theory.

At the heart of the problem is one undeniable truth: If an electron behaves like a wave, then something must be waving. But modern physics refuses to define what that something is.

Why Quantum Theory Refuses to Define What an Electron or a Photon Actually Is

Physics loves to classify things. We define the properties of mass, charge, and spin. We categorize subatomic particles into quarks, leptons, and bosons. We create elaborate quantum field theories to describe interactions.

But when it comes to what an electron or a photon actually is, physics has no answer. Is an electron a particle? A wave? A probability function? Is a photon a discrete particle of light? A disturbance in a field? A mathematical abstraction? If everything in quantum mechanics is both a wave and a particle, what does that actually mean?

Physics refuses to answer these questions because doing so would require rethinking the very foundations of the theory. If electrons are waves, then space cannot be empty—it must have structure. If photons are waves, then light does not need particles—just an underlying medium. If the double-slit experiment shows interference, then something physical must be oscillating.

Instead of addressing these contradictions, physics hides behind probability. Rather than admitting it doesn't know, it chooses to ignore the problem altogether.

This is where the Copenhagen Interpretation was born—the ultimate escape from responsibility.

The Copenhagen Interpretation: "Shut Up and Calculate" Instead of Questioning the Foundation

In the 1920s, when physicists realized that quantum mechanics made no sense, they had two choices. They could question their assumptions and search for a deeper reality. Or they could declare that reality doesn't exist until it is measured—and avoid the problem entirely.

They chose the latter.

The Copenhagen Interpretation, led by Niels Bohr, declared that particles do not have definite properties until they are observed. Reality was not an objective thing—only measurement made it real. Instead of searching for an underlying cause, physicists were told to "shut up and calculate."

This was the greatest intellectual failure in the history of physics. Instead of admitting that something was missing from quantum theory, they turned the lack of explanation into a feature. Instead of exploring why probability behaves like a physical wave, they declared that asking such questions was meaningless. Instead of searching for the real nature of space, waves, and particles, they invented a philosophy that avoided answering questions.

Quantum mechanics became a black box—a system that gives correct answers but refuses to reveal what is inside. We can calculate energy levels but don't know what energy actually is. We can predict outcomes but don't know what causes them. We can describe the behavior of particles but don't know what particles actually are.

Physics stopped being a search for truth and became an exercise in equations. This is why, a century later, quantum mechanics remains an unsolved paradox. We still don't know what waves in the double-slit experiment. We still don't know what an electron fundamentally is. We still don't know why measurement collapses a wave function.

Physics has made an entire industry out of ignoring these questions. Instead of seeking a real answer, it continues defending contradictions and calling it "progress."

But the truth is simple. Quantum mechanics is incomplete. Wave-particle duality is a mathematical fix, not a physical explanation. The missing piece is the true nature of space itself—a structured energy medium that physics refuses to recognize.

Next: The Fraud of Wave-Particle Duality – Light is Not Both a Wave and a Particle

If wave-particle duality is just a mathematical convenience, then what is really happening? Why did physics accept a contradiction instead of searching for a deeper explanation? Why do experiments clearly show wave behavior—but physics insists on calling it "particle-like"? How does modern physics reintroduce an invisible medium under different names—without admitting it?

In the next section, we tear down one of the biggest illusions in science—and replace it with the real answer. Because light, electrons, and all quantum interactions do not need contradictions—they need a proper physical explanation.

3.3 The Fraud of Wave-Particle Duality – Light is Not Both a Wave and a Particle

Physics has a habit of redefining reality to fit broken models rather than fixing the models themselves. One of the greatest examples of this is wave-particle duality—the idea that light, electrons, and other quantum entities somehow exist as both waves and particles at the same time.

This concept is presented as a fundamental truth of nature, but in reality, it is a convenient mathematical fix for a problem physics never truly solved. Instead of searching for the deeper structure of reality, physics accepted a contradiction. Instead of asking what light really is, physics divided it into two incompatible behaviors. Instead of recognizing that space itself may have an underlying structure, physics insisted that particles "sometimes act like waves."

Wave-particle duality was not a discovery—it was a forced compromise. And like all compromises in physics, it has led to a century of confusion, paradoxes, and unanswered questions.

It is time to expose why wave-particle duality is a fraud—and why light is not both a wave and a particle, but something else entirely.

How the Wave-Particle Duality "Solution" Was Just a Convenient Fix

The idea of wave-particle duality was never the result of an experiment proving that both states exist at once. In some experiments, light behaves like a wave—producing interference and diffraction. In others, it seems to act like a particle—such as in the photoelectric effect. Instead of questioning why these different behaviors emerge, physics simply declared that light is both a wave and a particle.

This solved nothing—it merely created an unnecessary paradox. If light is a wave, then what is waving? If light is a particle, then how does it interfere with itself? If light can switch between the two, what mechanism causes the switch?

Instead of addressing these contradictions, physics turned them into a rule: "Light is both a wave and a particle. Stop asking questions."

This is not science. This is a cover-up for a deeper misunderstanding of space, energy, and reality itself.

Why Physics Accepted a Contradiction Instead of Searching for a Deeper Explanation

Wave-particle duality is treated as an absolute fact today, but there was never a moment when physicists actually proved it. When quantum mechanics was still developing, scientists faced two competing models of light—wave theory and particle theory. Neither model worked perfectly, and experiments produced conflicting results.

Rather than admit that a deeper reality was missing, physics forced the two models together into an uneasy truce. Instead of fixing the mistake, they turned the mistake into a principle. Instead of asking what light actually is, they settled for equations that "worked." Instead of recognizing that waves require a medium, they created "virtual particles" to justify contradictions.

This is the same intellectual failure that led to space-time curvature and dark matter. Rather than rebuilding physics from a stronger foundation, scientists patched a broken model with mathematical tricks and paradoxical interpretations. And just like space-time curvature, wave-particle duality is an illusion designed to cover up what physics refuses to confront: the true nature of space and light.

How Experiments Clearly Show Wave Behavior—But Physics Insists on Calling It "Particle-Like"

Nearly every experiment in quantum mechanics shows overwhelming evidence that light behaves like a wave. Double-slit experiments create interference patterns, just like classical waves. Diffraction and polarization are wave phenomena that have no particle explanation. Photons spread out in space before being "measured," just like waves.

Yet, physics clings to the idea that light is also a particle—because of one experiment: the photoelectric effect.

This is the only reason photons exist as "particles" in modern physics. But even here, the wave explanation was never disproven. An energy wave interacting with electrons could explain the photoelectric effect just as well. But because space was assumed to be empty, waves were ruled out in favor of "particles."

The real mistake was assuming that light could only travel in a vacuum. If space were structured, light would not need photons—it would be a wave propagating through an energy field. If vacuum energy existed, light's behavior could be explained without wave-particle duality. If physics abandoned its obsession with particles, quantum mechanics would finally make sense.

Wave-particle duality is a desperate attempt to hold on to broken models. Light is not both a wave and a particle—it is a wave, and the "particle" interpretation is an illusion. The photon exists as a mathematical tool, not a physical entity. Experiments do not prove particles exist—they only prove that light transfers energy in quantized amounts.

Physics chose the wrong interpretation—and has spent a century doubling down on it.

The Missing Truth: Light is a Wave in a Structured Vacuum, Not a Photon Particle Hopping Through Space

If light is a wave, then what is waving? If space is truly empty, then nothing should be able to carry a wave. But if space has structure—an underlying energy field—then light is simply a wave moving through this medium. This eliminates the need for photons entirely.

Wave-particle duality exists only because physics refuses to admit the existence of a structured vacuum. The quantum vacuum already hints at energy fields permeating space. Gravitational waves suggest space itself has properties beyond mere geometry. The Higgs field is proof that energy fields exist everywhere, even in "empty" space.

Yet, physics refuses to connect these ideas—because doing so would mean admitting that photons are a mathematical convenience, not real objects. Light is not a stream of particles—it is a wave moving through an unseen energy field. Quantum mechanics is incomplete because it treats vacuum energy as mathematical rather than physical. Wave-particle duality is a symptom of a broken theory, not a fundamental truth of reality.

It is time to discard this century-old contradiction and search for the real answer.

Physics must stop defending paradoxes and start asking the real questions. If light is a wave, what is the medium? If the vacuum is structured, what are its properties? If space is not empty, how does it store and transfer energy?

The future of physics lies in answering these questions, not in preserving contradictions.

Next: The Real Nature of Light – How Physics Misinterpreted Its Own Experiments

If light is a wave in a structured vacuum, then why does physics still cling to photons? Why does modern physics secretly accept an invisible medium—but call it something else? How does quantum field theory contradict wave-particle duality but nobody talks about it? Why does space have structure—and why recognizing this fact will fix quantum mechanics forever?

In the next section, we expose the silent admission that physics is hiding. Because the truth has been there all along—and it's time to acknowledge it.

3.4 The Real Nature of Light: How Physics Misinterpreted Its Own Experiments

Physics has spent a century avoiding the obvious—light behaves like a wave because it is a wave. It interferes with itself. It diffracts around obstacles. It polarizes, spreads, and forms predictable patterns—just like any other wave.

Yet, modern physics insists that light must also be a particle, despite no direct evidence proving this. Why? Because accepting that light is purely a wave would force physics to admit that space is not empty. If light is a wave, then something must be waving. If space is truly a vacuum, then light should not propagate at all. If electromagnetic waves exist, then space must have an underlying structure.

Rather than confronting this reality, physics chose to misinterpret its own experiments. It introduced photons to explain energy quantization, even though a wave-based explanation was possible. It refused to accept the idea of structured space, even as quantum field theory reintroduced it under different names. It clung to the idea of an empty vacuum, even as experiments repeatedly suggested otherwise.

Now, the truth is emerging—light is a wave in a structured energy field, not a particle hopping through space. And modern physics has quietly admitted this—without saying it out loud.

Why Light's Wave Properties Prove That Space is Not Empty

Every classical wave—whether in air, water, or a solid medium—requires something to carry it. Sound waves travel through air. Water waves move through liquid. Seismic waves propagate through rock.

So what about electromagnetic waves? If space is truly empty, how do electromagnetic waves travel across it? If there is no medium, why does light have a constant speed? If photons were real particles, why do they still interfere and diffract like waves?

The only answer that makes sense is that space itself is not empty—it has structure. Light is not a particle—it is a wave propagating through this structure. The vacuum is not a void—it is an energy field that allows waves to exist. Space is not passive—it is active, dynamic, and fundamental to all interactions in physics.

Yet, despite overwhelming experimental evidence, physics refuses to state this directly. Instead, it has quietly reintroduced the idea of a structured vacuum—but under different names.

The Quantum Vacuum: A Hidden Structure Ignored in Mainstream Physics

In classical physics, the vacuum was assumed to be empty. Newtonian mechanics treated space as a featureless stage where objects moved. Einstein's relativity treated space as a geometric coordinate system with no substance. Quantum mechanics, however, began to reveal something different.

The more physicists studied quantum fields, the clearer it became: Space is not empty—it is filled with fluctuating energy. Virtual particles pop in and out of existence in this vacuum. The quantum vacuum affects real particles through the Casimir effect and zero-point energy.

Suddenly, the "empty vacuum" was no longer empty at all. It had measurable energy. It influenced particle behavior. It could even exert force.

Yet, physics still refused to call it a medium. Instead of admitting that space has structure, they called it "quantum fluctuations." Instead of recognizing it as the carrier of electromagnetic waves, they maintained the idea of photons. Instead of connecting vacuum energy to light propagation, they treated them as separate phenomena.

But the truth is unavoidable—if space has energy, then space itself is the medium for light. There is no need for photons. There is no need for wave-particle duality. There is only a structured energy field, carrying electromagnetic waves—just as air carries sound waves.

This is the real nature of light, and modern physics is slowly beginning to acknowledge it.

How Modern Physics Quietly Reintroduces an "Invisible Medium" But Refuses to Call It That

Even as mainstream physics insists that the vacuum is empty, it has been forced to introduce new ideas that contradict this assumption. Quantum field theory treats space as filled with fluctuating fields. The Higgs field is said to exist everywhere, giving particles mass. Dark energy fills space and drives cosmic expansion.

Each of these ideas suggests that space is not empty, but structured. Yet, physicists refuse to connect the dots. If space has energy, then it must also have properties. If space has properties, then it can act as a medium. If space is a medium, then electromagnetic waves do not need photons—they are simply waves in this medium.

Physics is quietly admitting this truth—but without challenging the old models. The idea of a structured vacuum is creeping back into physics—but under different names. Scientists speak of energy fields everywhere—but avoid calling them aether. They introduce dark energy and zero-point energy—but refuse to link them to the transmission of light.

This is the ultimate contradiction: Space is treated as empty for relativity. Space is treated as full for quantum mechanics. Space is ignored completely when it comes to wave propagation.

This is not a unified theory—it is a collection of contradictions. The only way to fix this is to acknowledge the truth—space itself is an energy-filled structure. Light is a wave in this structure. Gravity is an energy gradient in this structure. Quantum mechanics is incomplete because it refuses to define this structure.

It is time to abandon outdated concepts and embrace reality.

The Future of Physics: From Imaginary Photons to a Real, Structured Energy-Based Theory of Light

Physics has wasted a century defending paradoxes instead of searching for deeper truth. Wave-particle duality is not a real phenomenon—it is a symptom of an incomplete theory. Photons are not particles—they are a mathematical convenience. The vacuum is not empty—it is a structured energy field that carries light.

Once we recognize this, quantum mechanics will no longer be a mystery. Light will no longer need duality—it will be a wave, as it was always meant to be. Gravity will no longer be curvature—it will be an interaction with space's energy structure. Quantum entanglement, the double-slit experiment, and all other paradoxes will finally have real explanations.

The future of physics is not in defending broken models—it is in rewriting them from a foundation of truth.

Physics does not need more patches, contradictions, or mathematical tricks. It needs a new beginning—one that finally acknowledges the true nature of space, light, and energy.

And that is exactly what we will explore next.

Key Takeaways

- **Wave-particle duality** was introduced as a mathematical fix, not a true explanation of nature.
- Light was originally thought to be a wave, but the **photoelectric effect forced physics to introduce photons as particles.**
- Instead of questioning whether **space itself plays a role in wave behavior**, physics simply accepted the contradiction of duality.
- The **double-slit experiment** remains one of the biggest paradoxes in physics—why do single particles behave as waves?
- Quantum mechanics describes reality using **probabilities**, but it **does not explain what is actually happening beneath the wave function.**
- The acceptance of **wave-particle duality** stopped physics from investigating deeper mechanisms behind light and matter.

What If…? (Speculative Exploration)

- What if **light is not both a wave and a particle**, but just a structured energy wave interacting with the quantum vacuum?
- What if the **quantum wave function is not just a probability tool**, but a real, structured energy field?
- What if physics had explored **a structured vacuum instead of inventing photons**?
- What if the **photoelectric effect was not proof of particle behavior**, but an interaction with energy gradients in space?

Why Did Physics Accept the Contradiction of Wave-Particle Duality?

- The photoelectric effect **needed an explanation**, and the photon was an easy mathematical fix.

- The double-slit experiment **produced results that classical physics could not explain**, so quantum mechanics introduced probability waves instead of searching for a deeper mechanism.
- Instead of rethinking the nature of space and light, physics **forced light to be both a wave and a particle**.
- Mathematically, wave-particle duality "worked," so it was accepted without a proper physical explanation.

Challenges & Open Questions

- If light is not a particle, **why does it transfer energy in discrete packets?**
- If the quantum wave function is real, **what is it made of, and how does it interact with space?**
- How does **structured vacuum energy** explain wave behavior without requiring photons?
- What is **actually happening in the double-slit experiment**—is the electron truly interfering with itself, or is there an unseen energy structure guiding its motion?

Suggested Readings & References

- **De Broglie's Pilot-Wave Theory** – A hidden-variable approach that challenges wave-particle duality.
- **Tesla's Vision of a Structured Vacuum** – How electromagnetic waves interact with an energy-filled space.
- **Chavan's Universal Energy Dynamics**

A deeper explanation of wave behavior without paradoxes. (https://doi.org/10.5281/zenodo.14793162)

4: The Nuclear Force Myth

Fixing an Imaginary Problem

Physics has a long history of inventing forces to compensate for gaps in understanding. When gravity couldn't be explained, we curved space-time instead of questioning the nature of space itself. When light's behavior didn't fit classical physics, we forced it to be both a wave and a particle instead of rethinking our understanding of space and energy. And when physicists couldn't explain why atomic nuclei remain stable, they created the strong nuclear force—a force that has never been directly observed.

In previous chapters, we dismantled the illusion of space-time and exposed the contradiction of wave-particle duality. Now, we turn to another artificial fix in modern physics—the idea that atomic nuclei require a mysterious strong nuclear force to hold them together.

For decades, physicists have treated protons and neutrons as independent particles that require a separate force to keep them from flying apart. But what if this entire premise is flawed? What if protons and neutrons are not discrete billiard-ball-like particles needing a binding force? What if nuclear stability arises not from an invented force, but from an underlying energy structure? What if the strong nuclear force is just another mathematical patch, like dark matter and space-time curvature?

The moment physics assumed protons and neutrons were individual, self-contained particles, it created a problem that never existed. And like all such problems in physics, instead of questioning the assumption, scientists invented a new force to make the equations work.

This chapter exposes why the strong nuclear force is a myth and why the real explanation lies not in imaginary gluons, but in the structured vacuum energy that underlies all matter. The real reason the nucleus is stable. Why neutrons and protons are not bound by force, but by structured energy configurations. How discarding the strong nuclear force myth leads to a deeper understanding of atomic structure.

For over a century, physics has been trying to glue the nucleus together with an unnecessary force. Now, it's time to remove the glue and reveal the truth.

4.1 The Imaginary Glue: Why the Strong Nuclear Force Was Invented

One of the biggest myths in modern physics is that atomic nuclei require an invisible, all-powerful force to keep them from flying apart.

For decades, physicists have insisted that the strong nuclear force is a fundamental force of nature, responsible for binding protons and neutrons inside the nucleus. But this idea was never discovered—it was invented. It was a mathematical fix, not a physical truth. It was introduced because physics misunderstood the true nature of atomic structure. It became an unquestioned "law" despite never being directly observed.

To understand how this happened, we need to go back to the early days of atomic theory—when physicists made a fundamental assumption error that led them down a dead-end path.

The Historical Problem: Why Doesn't the Nucleus Fly Apart?

In the early 20th century, as scientists probed deeper into atomic structure, they encountered a major dilemma. The nucleus of an atom contains multiple protons, all of which have positive charges. Like charges repel each other—so, according to classical physics, the nucleus should explode due to electrostatic repulsion. Yet, nuclei remain stable—even in heavy elements like uranium, where dozens of protons are tightly packed together.

This posed a crisis. Either electromagnetism was incomplete, or something else was stabilizing the nucleus. Scientists assumed that an unknown force must be acting at the nuclear level to counteract electrostatic repulsion. Rather than questioning whether protons and neutrons were truly independent particles that needed a binding force, they simply invented one.

And just like that, the strong nuclear force was born. It wasn't discovered through observation. It wasn't proven through direct measurement. It was assumed to exist because the nucleus didn't behave as expected.

How Physics Assumed Protons Were Like Billiard Balls That Needed a Force to Hold Them Together

Physics has long been obsessed with reducing everything to individual particles. Atoms were modeled like miniature solar systems, with electrons orbiting a central nucleus. Protons and neutrons were assumed to be tiny, solid objects—like billiard balls, bouncing around inside the nucleus. If they were just sitting there as independent particles, then a separate force was needed to "glue" them together.

But this model never made sense. If the strong nuclear force was really as powerful as claimed, why doesn't it pull in nearby electrons? If protons and neutrons were truly separate particles inside the nucleus, how do they remain bound without orbiting or colliding? If the strong force acts only at nuclear distances, why does it "switch off" beyond a tiny range?

These contradictions were ignored because the strong nuclear force "solved" the problem—at least on paper.

The Birth of the Strong Nuclear Force – When Physics Invented a "Fix" Instead of Rethinking the Problem

Instead of reconsidering the nature of the nucleus itself, physics took the easy way out. It invented a new force. It introduced new, unobservable particles—gluons—to carry that force. It buried contradictions under quantum mechanics and called it progress.

But there was never direct evidence that:

- Gluons actually exist as real, independent particles.
- Protons and neutrons need a separate force instead of simply being stable energy structures.
- The nucleus holds together because of a "glue" rather than a deeper energy balance.

The strong nuclear force was not a discovery—it was a forced mathematical adjustment. It was created because the particle model of protons and neutrons required it. It was accepted because physics refused to question the flawed assumption behind it. It was locked into the Standard Model because nobody dared propose an alternative.

Why the Assumption That Neutrons and Protons Are Separate Particles Led to a Dead End

The biggest mistake in nuclear physics was treating protons and neutrons as independent particles needing a binding force. What if protons and neutrons are not "glued together" but instead are stable energy configurations? What if the nucleus doesn't need a binding force because it is an energy equilibrium, not a collection of objects needing glue? What if protons and neutrons are not separate physical entities but localized energy knots within a structured vacuum field?

These questions were never asked. Instead, physics forced the billiard-ball model to work by inventing a new force. Instead of considering that the vacuum itself plays a stabilizing role, physics created an entire framework of gluons and quark confinement. Instead of recognizing that the strong force was an unnecessary patch, physics built an entire industry around it.

But the cracks are starting to show. Quarks have never been observed as free particles, despite decades of high-energy experiments. Gluons remain hypothetical, only inferred through mathematical necessity. The nucleus behaves more like a stable energy system than a chaotic cluster of tiny particles.

These are signs that physics took a wrong turn.

The Real Question: Does the Nucleus Even Need a Binding Force?

If we discard the assumption that protons and neutrons are independent objects needing a glue-like force, we arrive at a much simpler answer. The nucleus is stable not because of an invented force but because of structured vacuum energy. Protons and neutrons are not separate particles but energy knots in this structure. The strong force is just another unnecessary fix—like space-time curvature and wave-particle duality.

Instead of treating the nucleus as a bunch of billiard balls stuck together, we must see it as an equilibrium of energy configurations. This explains nuclear stability without needing gluons or imaginary forces. This aligns nuclear physics with quantum field theory instead of contradicting it. This eliminates the last major patch holding up the Standard Model's illusion.

Physics never needed the strong nuclear force. What it needed was to recognize that its own assumptions were wrong.

Next: The Hidden Truth – The Nucleus Does Not Need a Binding Force

If the strong nuclear force isn't real, what keeps the nucleus together? How does structured vacuum energy naturally stabilize atomic nuclei? Why does the failure to observe free quarks prove that protons and neutrons are not what physics claims?

In the next section, we dismantle the illusion of the strong nuclear force and expose the real reason why the nucleus is stable. Because the truth is simple—the strong nuclear force was never needed in the first place.

4.2 The Hidden Truth: The Nucleus Does Not Need a Binding Force

For over a century, physics has assumed that atomic nuclei require a strong nuclear force to hold them together. This assumption was built on a flawed model—one that treated protons and neutrons as individual particles needing an invisible glue to keep them from flying apart.

But what if this assumption is completely wrong?

What if the nucleus is not held together by a separate force but by an energy equilibrium? What if nuclear stability arises naturally from structured vacuum energy, not an interaction of hypothetical gluons? What if the strong nuclear force is simply a misinterpretation of deeper energy dynamics?

Physics has spent decades forcing a complicated solution onto a problem that never existed. Now, it's time to dismantle the illusion and reveal the real reason why the nucleus is stable.

The Overlooked Possibility: What If the Nucleus Is Stable Because of Energy Balance, Not an Invisible Force?

Physics tells us that the nucleus consists of positively charged protons and neutral neutrons, bound together by the strong nuclear force. But this entire framework rests on a key assumption—that protons and neutrons are discrete, billiard-ball-like objects that require an external force to stay bound.

But what if this is completely wrong?

What if protons and neutrons are not separate objects but stable energy configurations? What if their binding is not due to an extra force but a balance of structured vacuum energy? What if the concept of nuclear forces is actually mistaking energy gradients for fundamental interactions?

Instead of questioning why the nucleus stays together, physics invented a force to explain what it didn't understand. But the truth is much simpler:

- The nucleus does not need a binding force—it exists in a stable energy state.
- The so-called strong force is an artifact of misunderstanding nuclear structure.
- The stability of the nucleus is a direct result of energy equilibrium, not a separate force.

Why Structured Vacuum Energy Can Naturally Stabilize Atomic Nuclei

Physics has long ignored a critical question—if space is truly a vacuum, why does it exhibit measurable energy effects?

Quantum mechanics has already revealed that space is not empty. Vacuum energy is everywhere, influencing particle behavior. The Higgs field exists throughout space, giving particles mass. The quantum vacuum produces measurable effects like the Casimir force.

Yet, physics refuses to consider that this structured vacuum energy might play a role in nuclear stability.

If vacuum energy interacts with particles, why wouldn't it also stabilize atomic nuclei? If quantum fields permeate all of space, why wouldn't they also dictate nuclear interactions? If forces emerge from energy gradients, why assume the nucleus requires a separate, fundamental force?

The strong nuclear force was only needed because physics ignored the role of structured vacuum energy.

Instead of seeing the nucleus as an energy equilibrium, physics treated it as a collection of particles requiring a force. Instead of recognizing the stabilizing effect of structured space, physics introduced gluons and confinement. Instead of acknowledging that vacuum energy could explain nuclear interactions, physics buried the idea under quantum chromodynamics.

But the real answer is simple:

- The nucleus is stable because it exists in a structured energy field—not because of an imaginary force.
- Protons and neutrons do not need to be glued together—they are stable energy knots within this structured space.
- The strong nuclear force is a patch, created to explain a problem that never existed.

The Illusion of Nuclear "Forces" – Mistaking Energy Gradients for Fundamental Interactions

One of the biggest mistakes in physics is confusing energy gradients with fundamental forces.

- **Gravity is not a force**—it is an energy gradient in structured space.
- **Electromagnetism is not a force**—it is an interaction between charged energy fields.
- **The strong nuclear force is not a force**—it is an emergent effect of structured vacuum energy stabilizing the nucleus.

Yet, physics continues treating these effects as separate fundamental forces instead of manifestations of underlying energy fields.

- The nucleus does not need a force to keep it together—it exists in a stable configuration due to energy balance.

- The concept of a "binding force" exists only because physics assumes that protons and neutrons are separate objects requiring glue.
- If physics had started from the assumption that space is structured, the strong nuclear force would never have been needed.

The belief in fundamental forces separate from structured space is a relic of outdated thinking.

Instead of recognizing that all interactions emerge from energy fields, physics invented separate, disconnected forces. Instead of seeing nuclear stability as an equilibrium, physics forced an unnecessary force into existence. Instead of unifying physics through a structured energy model, scientists created an ever-growing list of particles and forces.

The strong nuclear force is just another imaginary fix—one that collapses once we recognize the stabilizing role of vacuum energy.

How the Failure to Detect Free Quarks Proves That Protons and Neutrons Are Not What Physics Claims

The Standard Model claims that protons and neutrons are made of quarks, held together by gluons. But there is a problem—free quarks have never been observed.

For decades, physicists have tried to isolate individual quarks using high-energy collisions, but have failed every time.

If quarks were real, why can't they exist as separate particles? If gluons "hold quarks together," why does their binding mechanism behave unlike any other force? If the strong force is fundamental, why does it seem to "switch off" at larger distances?

These contradictions should have forced physics to rethink its assumptions. But instead, physics invented another fix—quark confinement.

Instead of asking why quarks never appear alone, physics declared that they cannot be separated. Instead of reconsidering the structure of protons and neutrons, physics created color charge and asymptotic freedom. Instead of recognizing that protons and neutrons are stable energy knots, physics added complexity to preserve the Standard Model.

But the truth is simple:

- Quarks have never been observed as free particles because they do not exist as separate objects.
- Protons and neutrons are not collections of smaller particles—they are stable energy configurations.
- The nucleus is not held together by an imaginary force—it is stabilized by energy equilibrium in structured vacuum energy.

Next: Neutrons and Protons – Energy Knots, Not Force-Bound Particles

- If neutrons and protons are not held together by the strong force, what are they?
- Why does energy behave like a "particle" at nuclear scales?
- How does structured space naturally produce the stability observed in atomic nuclei?

In the next section, we expose the true nature of nuclear particles—and why physics has misinterpreted them for a century. Because once we discard the illusion of the strong nuclear force, the true nature of atomic structure finally becomes clear.

4.3 Neutrons and Protons: Energy Knots, Not Force-Bound Particles

Physics has spent the last century treating protons and neutrons as if they were independent, billiard-ball-like particles that need a mysterious force to hold them together.

This assumption led to the invention of the strong nuclear force. It also led to the creation of quarks and gluons—hypothetical particles that were never directly observed. And it locked nuclear physics into an unnecessarily complicated framework that ignores the possibility of a structured vacuum.

But what if this entire particle-based approach is wrong?

What if protons and neutrons are not individual particles at all? What if they are localized energy configurations, naturally stabilized by the structure of space itself? What if the strong nuclear force is unnecessary because the nucleus is already in an energy equilibrium?

This section dismantles the illusion of force-bound particles and reveals the real nature of nuclear matter.

Why Treating Protons and Neutrons as Separate, Self-Contained Particles Is a Mistake

Since the early 20th century, physicists have assumed that protons and neutrons are fundamental nuclear particles—tiny, self-contained objects that exist independently and require a force to hold them together inside the nucleus.

This assumption created the need for an extra force—the strong nuclear force. It led to the concept of quarks—substructures that have never been observed in isolation. It forced physics to invent gluons—particles that exist only as a mathematical requirement, not as direct experimental evidence.

But there's a fundamental flaw in this assumption.

- Protons and neutrons are never truly observed outside an atomic system.

- Free neutrons decay within minutes, which means they are not stable standalone objects.
- Protons exhibit behavior that suggests they are more than just singular, self-contained entities.

So what if protons and neutrons are not particles in the way physics describes them?

What if they are energy formations—localized knots of energy in a structured vacuum? What if they do not need a separate binding force because their structure is already stable in an energy equilibrium? What if nuclear stability is not the result of a force but the result of an underlying energy structure?

This shifts the entire perspective of nuclear physics.

Instead of particles needing glue, we see energy forms naturally holding their shape. Instead of a strong nuclear force, we see a balance of structured vacuum energy. Instead of disconnected particles, we see a unified energy system at the nuclear level.

Once we abandon the billiard-ball model of protons and neutrons, the need for the strong nuclear force disappears.

How Neutrons and Protons Are Localized Energy Knots, Not Discrete Objects Requiring a Glue-Like Force

Every stable structure in nature follows one rule—balance is more fundamental than force.

From atoms to galaxies, stability emerges not from glue-like forces but from equilibrium in an underlying energy structure. So why should the nucleus be any different?

If space is structured, then nuclear particles should emerge as stable formations within that structure. If vacuum energy is real, then protons and neutrons should be energy knots within that vacuum, not isolated objects. If nature prefers stability, then the nucleus should not need an artificial force—it should naturally maintain its form.

This means:

- Protons and neutrons are not "stuck together" by a force—they exist as stable, co-existing energy patterns.
- They do not require gluons—they exist in a vacuum energy equilibrium.
- They do not switch between quark states—they are unified structures shaped by energy density variations in space.

This eliminates the need for a strong force altogether.

- What physics calls a "force" is just the appearance of energy balance.
- What physicists call "quark interactions" are just internal redistributions of energy within a localized energy knot.
- What physics calls "binding" is nothing more than stable configurations of structured energy.

The Failure of the Quark-Gluon Model – Why Nobody Has Ever Observed Quarks as Independent Particles

The Standard Model tells us that protons and neutrons are made of quarks, held together by gluons. But there's a serious problem—no one has ever observed a free quark.

Despite decades of high-energy experiments, quarks never appear alone. Despite smashing protons apart in particle accelerators, quarks never emerge independently. Despite increasing energy levels, the so-called strong force seems to get stronger, never allowing quarks to separate.

This led to an absurd conclusion—quarks must be confined.

Physics has never directly observed quarks, but instead of reconsidering the model, scientists created an excuse: *quark confinement.*

They claimed quarks can never exist freely because the strong force prevents them from separating. They invented "color charge" to explain why quarks behave differently from other particles. They introduced asymptotic freedom—a property that conveniently explains why quarks seem free inside protons but never outside them.

But what if quarks are missing not because they are confined, but because they were never separate objects to begin with?

- What if protons and neutrons are not "made of" quarks but are just stable energy formations?
- What if the so-called "quark interactions" are just shifts in the energy density of a single, unified structure?
- What if the nucleus works not because of internal particle forces but because of an external vacuum energy structure?

Once again, the assumption that matter is made of particles leads physics to create an unnecessary force.

The strong force exists only because physics refuses to abandon the quark-gluon model. Quark confinement exists only because no one wants to admit quarks may not be real particles. The complexity of nuclear physics exists only because scientists refuse to consider structured space as the real answer.

It's time to discard this flawed model and recognize nuclear particles for what they really are—localized energy knots in a structured vacuum.

Next: Rethinking Atomic Stability – From Nuclear Forces to Energy Equilibrium

- How structured vacuum energy explains why nuclei remain stable.
- Why adding more particles and forces has only complicated the problem.
- How physics can move from an outdated particle-force model to a true energy-based understanding.

In the next section, we complete the dismantling of nuclear forces and reveal the final missing piece. Because the strong force was never needed—all we needed was to recognize the true structure of space.

4.4 Rethinking Atomic Stability: From Nuclear Forces to Energy Equilibrium

For over a century, nuclear physics has been built on a patchwork of imaginary forces, invented particles, and mathematical constructs—all designed to compensate for a fundamental misunderstanding of atomic structure. When physicists couldn't explain why the nucleus stays together, they invented the strong nuclear force. When protons and neutrons didn't behave like fundamental particles, they introduced quarks. When quarks couldn't be observed, they created gluons and "confinement" to justify their absence.

At every step, instead of questioning the foundation, physics added more complexity. But the real answer has been hiding in plain sight. The nucleus is stable not because of an invisible force, but because of structured energy equilibrium. Protons and neutrons are not independent particles needing a "glue"—they are stable energy knots within a structured vacuum. The so-called strong nuclear force is just another misinterpretation of deeper energy dynamics.

This final section completes the dismantling of nuclear forces and presents the real reason why atomic stability exists—not because of separate forces, but because of the fundamental nature of space itself.

Why Adding More Particles (Quarks, Gluons, Mesons) Only Complicates the Problem Instead of Solving It

Physics has a bad habit: whenever a theory fails, instead of questioning the assumptions, it adds more particles. The proton and neutron weren't enough, so quarks were introduced. When quarks didn't appear in experiments, physicists invented "confinement" and gluons to explain their absence. When that still wasn't enough, mesons and virtual particles were added.

At every stage, rather than admitting that protons and neutrons might not be what we assumed, physics doubled down on the mistake. The quark-gluon model became a mathematical necessity, not an experimental discovery. The Standard Model expanded into an increasingly bloated framework filled with unproven entities. The complexity grew, but the contradictions remained.

This is not progress—it's a desperate attempt to fix a problem that doesn't exist. If nuclear forces were real, why do they behave so differently from other fundamental forces? If quarks were real, why has no one ever observed them as independent particles? If the Standard Model was truly complete, why does it require constant adjustments to make predictions work?

The problem isn't that we need more particles—the problem is that we started with the wrong assumptions. Discarding the idea that protons and neutrons are separate objects needing a binding force eliminates this unnecessary complexity. Instead of imaginary particles and forces, a unified energy model is needed—one that explains both nuclear and gravitational stability.

The Missing Link: How the Same Energy Structure That Explains Gravity Also Explains Nuclear Stability

In previous chapters, we dismantled space-time and revealed that gravity is not a force but an effect of energy gradients in structured vacuum energy. Why should nuclear stability be any different? If gravity emerges from energy density variations in space, why not nuclear interactions? If mass is a localized energy formation, why wouldn't the nucleus be structured by the same principles? If space itself has an underlying energy structure, why wouldn't it govern all fundamental interactions—including atomic stability?

Physics has always treated gravity and nuclear forces as separate realms, when in reality, they are just two different scales of energy equilibrium within the same structured vacuum. What we call "gravity" is simply the large-scale effect of energy gradients. What we call "nuclear binding" is the small-scale effect of localized energy balance. Neither requires a separate force—both emerge naturally from structured space.

This realization changes everything. Gravity and nuclear stability are not governed by different fundamental forces, but by the same underlying energy dynamics. The moment we recognize that all forces emerge from structured energy, the Standard Model collapses into a much simpler and unified framework.

The Future of Physics: From Imaginary Forces to a Unified Energy-Based Model

The current model of nuclear physics is unsustainable. It is filled with ad hoc fixes, relies on particles that have never been directly observed, and contradicts the principles of quantum mechanics while refusing to unify with gravity. The only way forward is to abandon the force-based model and embrace an energy-based understanding of reality.

Discarding the idea of quarks and gluons as fundamental particles reveals them as mathematical constructs, not physical entities. Stop treating protons and neutrons as separate objects needing a binding force—they are energy formations that stabilize naturally. Acknowledge that nuclear stability is an effect of structured vacuum energy, just like gravity.

This shift leads to a completely new way of understanding fundamental interactions. No more artificial "fundamental forces." No more unnecessary complexity in the Standard Model. A unified model where all interactions are energy-based, not force-based.

Physics has spent a century defending an outdated view of nature—one where forces act separately instead of emerging from a unified structure. It is time to correct this mistake and rebuild physics on a foundation of reality.

How Discarding the Strong Nuclear Force Myth Leads to a Deeper Understanding of Mass, Charge, and Fundamental Interactions

Accepting an energy-based nuclear model forces physics to rethink mass, charge, and the true nature of interactions. Mass is not a fundamental property of particles—it is an effect of energy density in structured space. Charge is not a mysterious intrinsic property—it is a dynamic interaction between energy fields. Forces do not exist as independent entities—they are all emergent effects of energy gradients.

This transformation unifies gravity and nuclear interactions under a single framework. It removes unnecessary layers of complexity in particle physics. It opens the door to a new era of physics—one where we finally stop adding imaginary fixes and start uncovering real fundamental truths.

Physics does not need more particles. It needs a complete rethinking of its assumptions. The Standard Model, with its endless particles and forces, is a temporary solution to a misunderstanding. The moment we accept that all interactions are energy-driven, we no longer need to force separate rules for nuclear physics and gravity. The strong nuclear force is just another illusion that collapses when we recognize the true nature of space and energy.

The future of physics will not be about discovering more particles—it will be about understanding the structure of energy itself. Because once we do, we will finally see the universe as it truly is.

Key Takeaways

- The **nuclear force** was introduced as a mathematical fix to explain why protons in the nucleus do not repel each other due to their positive charge.
- Instead of questioning whether **electromagnetic repulsion is the only force at play**, physics **invented the strong nuclear force** to hold protons together.
- Quantum chromodynamics (QCD) introduced **quarks and gluons**, making the model even more complex rather than providing a fundamental explanation.
- The concept of **binding energy** was misunderstood—energy does not "glue" particles together but arises from a deeper structural arrangement.
- **Energy structuring at a fundamental level** could explain nuclear stability **without needing an arbitrary strong force**.

What If...? (Speculative Exploration)

- What if the **strong nuclear force does not exist**, and nuclear stability is a result of structured energy fields?
- What if protons and neutrons are not **held together by a mysterious force**, but by a balance of fundamental energy configurations?
- What if quarks and gluons are **just mathematical constructs**, rather than actual physical entities?
- What if nuclear reactions and fission are **better explained by energy redistribution** rather than a "force" holding the nucleus together?

Why Did Physics Create the Strong Nuclear Force?

- **Protons should repel each other**, but atomic nuclei remain stable, so physics needed a new force to explain this contradiction.
- Instead of reconsidering **how charge and energy interact at a fundamental level**, physics **patched** the problem with an arbitrary force.
- **Gluons and quarks** were added to make sense of nuclear interactions, but **none of them have been observed directly**.
- Quantum chromodynamics (QCD) became **so mathematically complex** that physicists focused on making calculations work rather than questioning its validity.

Challenges & Open Questions

- If the strong nuclear force does not exist, **how can we explain the stability of atomic nuclei?**
- Could an **energy-based approach** provide a more natural explanation for nuclear interactions?
- Is it possible that nuclear forces are just **electromagnetic interactions behaving differently at extreme energy densities**?
- Could structured vacuum energy explain **why protons and neutrons arrange themselves the way they do**?

Suggested Readings & References

- **Electromagnetic Theories of Nuclear Stability** – Alternative perspectives on nuclear interactions.
- **Tesla's View on Energy and Matter** – Why forces may not be fundamental, but emergent from structured energy.
- **Chavan's Energy Based Atomic Configuration** – A unified approach to explaining atomic stability without a strong nuclear force.

5: The Standard Model Disaster
A Frankenstein of Fixes

For decades, the Standard Model has been hailed as the pinnacle of human understanding in particle physics. It is praised as "the most successful theory ever formulated," a framework that supposedly accounts for all fundamental forces—except gravity. It boasts remarkable experimental confirmations, including the celebrated discovery of the Higgs boson.

But beneath the layers of accolades lies an inconvenient truth. The Standard Model is not a revelation of nature—it is a desperate assembly of mathematical patches. It does not unify physics—it merely adds more components to compensate for contradictions. It does not explain reality—it forces reality to conform to outdated equations that should have been discarded long ago.

In the previous chapters, we dismantled the illusions of fundamental forces, showing how nuclear stability and gravity arise from structured energy fields rather than separate force interactions. Now, we turn to the biggest illusion of all—the Standard Model itself. How it was cobbled together to keep broken theories afloat. Why quarks, gluons, and bosons were "discovered" not because they exist, but because equations demanded them. Why the endless cycle of adding new particles will never lead to a fundamental theory—because the foundation is already flawed.

The Standard Model was not built from an honest search for truth—it was constructed to hide the unresolved contradictions of particle physics. Every "missing piece" led to another forced invention. Every inconsistency required another layer of complexity. No amount of discoveries will ever complete it—because it was broken from the very beginning.

Physics has spent a century celebrating an illusion. It's time to tear apart the Frankenstein's monster we've been calling a theory.

The Patchwork Model: A Theory That Grows Instead of Simplifies

A real fundamental theory should unify and simplify our understanding of nature. It should reduce complexity, not increase it. Yet, the Standard Model has done the opposite. Instead of revealing a deeper, unified reality, it has grown into a monstrous collection of ad hoc fixes and unexplained assumptions.

The theory began as an attempt to explain particle interactions, but every time an observation didn't fit, new particles were added to compensate. The discovery of the Higgs boson was treated as the crowning achievement of the model, but in reality, it was just another forced addition—introduced not because a field was directly observed, but because the equations required it to "exist."

At every stage, instead of questioning whether the framework itself was flawed, physics took the easy way out—expanding the model rather than simplifying it. Instead of identifying a deeper, unified principle, it created a sprawling network of force carriers, virtual particles, and interaction rules that defy any natural logic.

What kind of "fundamental" theory constantly needs to be modified to remain valid?

Why the "Discovery" of Quarks, Gluons, and Bosons Was Driven by Equations—Not Experiment

The Standard Model tells us that everything is made of smaller building blocks: quarks, held together by gluons, exchanging forces through bosons. But there's a problem—none of these so-called fundamental particles exist as independent, observable entities.

Quarks were never actually seen in isolation. Their existence was inferred because high-energy collisions produced particle jets that could be mathematically explained by assuming smaller components inside protons and neutrons. Instead of considering alternative explanations, physics simply accepted quarks as real because they made the equations work.

The same thing happened with gluons. They were never directly observed. Their presence was inferred through indirect energy shifts. And yet, they were immediately enshrined as real force-carrying particles because the Standard Model required them.

The Higgs boson followed the same pattern. Instead of questioning whether mass might arise from structured space rather than a particle interaction, physicists built enormous experiments to confirm the existence of the Higgs—not because they had direct evidence it was real, but because the Standard Model demanded it. When a particle-like signal appeared at the expected energy range, it was declared a triumph. But the entire logic behind the Higgs field remains unchallenged.

This is how modern physics works: If an equation needs a particle, experiments are designed to find something that fits the prediction, and once any anomaly appears near the expected energy levels, it is labeled as a "discovery."

The Standard Model is not a result of discovering nature—it is a result of forcing nature to fit equations.

A Theory That Can Never Be Completed

The greatest irony of the Standard Model is that it will never be finished. Every time an inconsistency appears, a new particle is added. Every time a measurement deviates from the model, a "new physics" explanation is proposed. The cycle never ends because the foundation itself is unstable.

If the model were truly fundamental, it would explain gravity, but it doesn't. If it were truly complete, it would require no additional components, yet dark matter and dark energy remain mysteries. If it were truly accurate, it wouldn't require constant modifications to maintain consistency with experiments.

The Standard Model is not a final theory—it is a never-ending work in progress, a collection of forced solutions that grow more tangled with every new discovery. It doesn't explain nature; it merely provides a set of mathematical rules that describe observations well enough to avoid admitting that something is missing.

The Future of Physics: Tearing Down the Standard Model and Rebuilding from Reality

Physics has reached a dead end. The obsession with force-based interactions and particle-based explanations has led to a theoretical framework that is bloated, untestable, and fundamentally detached from reality. The only way forward is to abandon the force-and-particle model and embrace a structured energy-based understanding of the universe.

Instead of treating particles as the foundation of matter, we must recognize that they are emergent effects of energy configurations in structured space. Instead of force carriers like bosons and gluons, we must see interactions as energy gradients rather than particle exchanges. Instead of searching for ever-smaller building blocks, we must accept that structure itself—not smaller and smaller particles—is the key to understanding nature.

This shift requires letting go of the Standard Model's artificial complexity. It requires recognizing that mass, charge, and fundamental forces are all emergent properties of structured energy fields rather than separate interactions mediated by particles. It requires redefining our understanding of reality from the ground up.

Physics does not need more particles. It does not need more layers of theoretical fixes. It needs a fresh start, free from the mathematical traps that have turned it into a self-referential maze.

The Standard Model is not a masterpiece—it is a mistake. And until physics admits this, it will remain stuck, chasing ghosts instead of seeking truth.

5.1 The Patchwork Theory: How the Standard Model Was Forced to Fit Broken Theories

Physics was once a quest to uncover the fundamental nature of reality. But somewhere along the way, it became an exercise in salvaging broken theories with mathematical patches. No example illustrates this better than the Standard Model of particle physics—a framework that was meant to unify forces and explain the nature of matter but instead became a growing collection of forced fixes.

The Standard Model was supposed to describe nature, not force nature to conform to equations. It was expected to be the final answer to particle physics, yet it has evolved into a Frankenstein of contradictions—an ever-expanding list of particles and forces, each invented to plug the gaps in an incomplete theory.

The Original Goal of the Standard Model – A Unified Theory of Forces

By the early 20th century, physics was at a crossroads. Einstein had unified space and time in relativity, but gravity refused to fit within quantum mechanics. Quantum mechanics described atomic behavior but lacked a deeper explanation for why particles behaved as they did. The discovery of nuclear interactions demanded a new theory beyond electromagnetism and gravity.

The Standard Model aimed to bring all known particles and forces under one framework. Initially, it seemed promising. Physicists organized particles using group theory and quantum field equations, successfully describing electromagnetism and weak interactions. Predictions like the W and Z bosons were later experimentally confirmed, strengthening confidence in the model.

But cracks soon appeared. When experimental results didn't match predictions, instead of questioning the model, physics simply added another particle to make it work. Each contradiction led to another layer of complexity. Every unexplained phenomenon became an excuse to introduce something new rather than re-evaluating the foundation.

The result? A model that works mathematically but lacks physical meaning. A theory that only "predicts" particles after they are required to fix inconsistencies. A framework so bloated that it no longer explains nature—it forces nature into compliance. And the cycle never stopped.

Every Contradiction Led to a New Particle Instead of a New Approach

The biggest failure of the Standard Model is its reliance on new particles as fixes rather than rethinking fundamental assumptions. Instead of refining its foundations, physics took the easy route:

- Protons and neutrons weren't fundamental? Invent quarks.
- Quarks needed to be held together? Invent gluons.
- Electroweak unification required mass? Invent the Higgs boson.
- Particles had unexpected decay patterns? Invent flavors of neutrinos.
- Matter-antimatter asymmetry remained unexplained? Invent CP violation and additional quantum numbers.

At no point did physics stop to ask:

Is the particle-based model even correct? Does reality truly function through discrete force interactions? Could energy structures explain these effects instead of new particles?

Instead of reconsidering whether protons and neutrons are even made of smaller components, physics created quarks—despite never observing them directly. Instead of recognizing that nuclear stability might arise from structured energy rather than a separate force, physics introduced gluons to justify an unnecessary binding force. The Higgs boson was proposed not because of direct evidence, but because the equations required it to exist. Dark matter was introduced simply because the Standard Model failed to explain galaxy motion.

None of these so-called "discoveries" were genuine breakthroughs. They were desperate attempts to keep an outdated model alive.

The Never-Ending Cycle of Patches – Why Every "Discovery" Adds More Complexity

A real scientific theory should predict observations before they happen, not after. It should become simpler over time, not increasingly complex. It should explain why things happen, not just describe how they appear.

The Standard Model fails on all three counts.

The Higgs boson was not discovered as a natural outcome of theory—it was postulated long before any experimental evidence appeared, simply because the equations demanded it. The top quark was inserted into the framework to fill a mass gap in particle families. Neutrino oscillations forced physics to modify assumptions about particle interactions.

Each time physics found an inconsistency, instead of rethinking the foundation, another patch was added:

- The strong force behaves differently from other forces? Invent "asymptotic freedom."
- Quarks never appear freely? Invent "quark confinement."
- The Higgs boson doesn't behave like a normal particle? Invent "vacuum expectation values."

This is not how real science works. It is a sign of a failing theory held together by ever-increasing complexity.

The Elephant in the Room: If the Standard Model Is So "Successful," Why Does It Need Constant Adjustments?

Physicists insist that the Standard Model is "the most successful theory ever created." They claim it explains everything—except gravity, dark matter, dark energy, quantum gravity, and unification. They say it is complete—yet keep adding new particles to fix inconsistencies. They celebrate its accuracy—but only after modifying equations to match experimental results.

If the Standard Model is truly successful, why does it require:

- Supersymmetry—a completely new theoretical framework—to "fix" its missing pieces?
- Dark matter—a completely separate explanation—to account for the universe's mass?
- Constant adjustments to particle interactions every time experiments disagree with predictions?

The truth is simple:

The Standard Model is not successful—it is just convenient. It provides enough statistical accuracy to keep physicists invested. It delivers just enough particle discoveries to justify funding. It ensures physics remains locked in an endless cycle of "discoveries" that never actually solve anything.

It is not a fundamental theory of nature. It is a continuous experiment in forcing equations to fit observations, no matter how absurd those equations become.

Next: The Myth of Quarks, Gluons, and Bosons – Why These Particles Don't Really Exist

- Why quarks were never actually discovered—only inferred through indirect evidence.
- How gluons were mathematically required but have no independent existence.
- Why bosons, including the Higgs, are convenient fixes rather than fundamental particles.
- How quantum field theory contradicts the very existence of individual force carriers.

In the next section, we dismantle the biggest illusion in particle physics—the idea that quarks, gluons, and bosons exist as real physical entities. Because once we discard this illusion, we can finally begin to understand the true nature of matter and energy.

5.2 The Myth of Quarks, Gluons, and Bosons: Why These Particles Don't Really Exist

The Standard Model claims to break down all matter and forces into fundamental particles. It insists that protons and neutrons are made of quarks, held together by gluons, while fundamental forces are carried by bosons. This narrative is presented as settled science, yet there's a glaring problem—none of these particles have ever been observed in isolation.

No one has ever directly seen a quark. No experiment has ever isolated a gluon. The so-called "discovery" of the Higgs boson was not a direct detection but a statistical inference. Despite this, these entities are still promoted as real, even though they exist only as mathematical constructs—introduced out of necessity to make the equations work.

The belief in quarks, gluons, and bosons is not based on physical evidence but on an effort to patch an incomplete theory. Their existence was never confirmed—it was assumed because the model required them. This is not how fundamental physics should work.

The Invention of Quarks—A Mathematical Fix That Was Never Observed

Before the quark model, protons and neutrons were considered fundamental particles. They had measurable mass, charge, and spin, interacted through known forces, and could be directly observed. But as high-energy experiments probed deeper, physicists noticed that protons and neutrons seemed to have internal structure when bombarded with electrons. Their scattering patterns suggested they might be composed of even smaller constituents.

Rather than reconsidering the nature of nuclear energy or exploring alternative explanations, physics took the simplest route—it assumed protons and neutrons must be made of smaller "building blocks" and named them quarks. But there was a problem:

No quark has ever been directly observed. No experiment has isolated a free quark. The explanation? Physics declared quarks are permanently "confined" and can never exist independently. This is not a discovery—it is an excuse.

Every other so-called "fundamental" particle in physics can exist on its own—so why not quarks? If they are real, why do they always appear in groups, never individually? The simplest answer is that quarks are not real particles at all. They are mathematical placeholders used to describe patterns in experimental data, not physical objects that exist in nature. Their "existence" was inferred, not observed. They were introduced to make the Standard Model function, not because they were found in nature.

But quarks alone weren't enough. If they were real, something had to hold them together. And so, physics invented another imaginary particle.

The Creation of Gluons—A Force Carrier That No One Has Ever Isolated

Once quarks were introduced, physicists needed to explain why they stayed bound inside protons and neutrons. They couldn't use electromagnetism, because quarks supposedly had a different type of charge called "color charge." Gravity was too weak to explain the binding. The strong nuclear force only applied to protons and neutrons as a whole, not their internal components.

Rather than questioning whether quarks were real, physics took the next step in patching the model—it invented gluons. These were defined as massless particles that "bind" quarks together, providing the strong interaction that prevents them from flying apart. But, just like quarks, gluons have never been observed in isolation. Their existence is assumed based on the behavior of high-energy collisions, but they are never directly measurable.

This creates an unsolvable contradiction. If quarks are real, they should be observable as individual particles. If gluons are real, they should be detectable. But instead, physics claims that both are forever hidden by the very force that allegedly binds them together. A theory that continually explains away the absence of its predictions is not science—it is an artificial construct designed to preserve a broken model.

Instead of questioning whether protons and neutrons are actually made of quarks, physics doubled down. It assumed quarks were real, assumed they needed a force carrier, and created gluons simply because the equations required them. But the Standard Model's reliance on imaginary particles didn't stop there.

The Problem with Bosons—Why the Higgs, W, and Z Bosons Are Just Mathematical Placeholders

Once physics accepted that forces needed "carrier particles," it became a never-ending cycle of particle discoveries. The electromagnetic force was assigned a carrier—the photon. The weak nuclear force required additional carriers—W and Z bosons. The Higgs boson was introduced because equations required a mechanism to "give" particles mass.

These particles were not discovered first. They were proposed out of mathematical necessity and later "found" in high-energy collisions, where statistical probabilities—not direct detection—determined their existence. The Higgs boson, for example, was declared discovered not because a distinct particle was seen, but because decay patterns in experiments slightly deviated from background noise in a way that matched theoretical predictions.

Instead of simplifying physics, each of these particles only made the model more complicated. If the Higgs field gives mass, why do particles have different masses? If the weak force has massive carriers, why does it behave so differently from electromagnetism? If bosons exist as force carriers, why do they behave like particles rather than continuous fields?

These are fundamental contradictions that the Standard Model refuses to address. Rather than questioning whether forces truly require discrete particles, physics keeps adding new ones to patch the gaps.

Quantum Field Theory Itself Contradicts the Existence of Force-Carrying Particles

The biggest irony of modern physics is that quantum field theory, the very framework that underlies the Standard Model, contradicts the idea of discrete force carriers. Fields are continuous structures, not made of particles. Forces emerge from interactions in these fields, not from the "exchange" of tiny messenger particles.

This means that the very concept of force-carrying particles is an unnecessary oversimplification. Forces are not transmitted by discrete particles—they emerge from structured energy fields. Mass is not "given" by the Higgs boson—it results from energy equilibrium within the structure of space itself.

Once physics acknowledges this, the entire Standard Model collapses. Quarks, gluons, and bosons are unnecessary. The illusion of fundamental force carriers disappears. And the search for ever more "missing particles" to fix contradictions finally comes to an end.

Next: The Illusion of Completeness – Why Adding More Particles Won't Solve Anything

Physics continues to invent "missing particles" rather than addressing its flawed foundation. The search for dark matter, sterile neutrinos, and axions is a symptom of a broken model, not a path to deeper understanding. Instead of endlessly adding more particles to fix contradictions, it is time to move beyond the outdated particle-based model and recognize that reality is structured by energy, not discrete building blocks.

The only way forward is to abandon the illusion and start asking the right questions. Because once we stop forcing nature to fit our equations, we can finally begin to understand the universe as it truly is.

5.3 The Illusion of Completeness: Why Adding More Particles Won't Solve Anything

The Standard Model is often celebrated as one of the greatest achievements in modern physics. It claims to explain nearly all known particles and forces, requiring only minor adjustments. It supposedly predicts experimental outcomes with extraordinary precision. Yet, despite these grand claims, the model remains perpetually incomplete.

If the Standard Model were truly a finished theory, why does it constantly demand new additions? Why must physics still chase after axions, sterile neutrinos, and supersymmetric particles? Why do dark matter and dark energy remain mysteries, with no viable explanation within the framework?

Every so-called "discovery" seems to generate more unknowns, leading to yet another round of theoretical adjustments.

The reason is simple: the Standard Model is not complete. It is a self-sustaining system that creates new gaps every time it tries to patch an old one. Instead of questioning its flawed assumptions, physics continues to add particles as quick fixes, avoiding the real problem—the model itself.

How Every "Missing Piece" of the Standard Model Creates New Gaps That Require Even More Particles

For decades, physicists have promised that the Standard Model was almost complete, just a few missing pieces away from perfection. The discovery of the Higgs boson was supposed to mark the final triumph, yet it only raised new questions. Supersymmetry was introduced to fix contradictions in quantum mechanics, but it now predicts an entire zoo of undiscovered particles. Dark matter was expected to be a straightforward addition, yet it has spawned endless speculation about exotic particles with no experimental confirmation.

Every time one gap is "solved," another appears. Neutrinos were originally thought to be massless, but oscillations were discovered, forcing new modifications. Quarks were introduced as simple components of protons and neutrons, but their supposed behavior demanded the addition of color charge, confinement, and increasingly complex interactions. The Higgs boson was meant to explain mass, but its own mass remains an unsolved mystery, requiring yet another theoretical adjustment.

Instead of recognizing that the Standard Model is built on shaky foundations, physicists continue to pile on more particles, more forces, and more equations to keep the structure from collapsing. But a real theory simplifies nature—it does not make it infinitely more complicated.

Why Physics Keeps Adding Axions, Sterile Neutrinos, Supersymmetric Particles, and Dark Matter Candidates Instead of Questioning Their Necessity

Modern physics has become less about searching for truth and more about searching for particles. Axions were introduced to fix a mathematical problem in quantum chromodynamics, but they are now treated as a leading candidate for dark matter. Sterile neutrinos were suggested to explain neutrino oscillations, yet they have evolved into an explanation for missing cosmic mass. Supersymmetric particles were originally meant to stabilize quantum mechanics, but now they are invoked to explain unification, dark matter, and even cosmic inflation.

The pattern is obvious. Every time physics fails to explain an observation, a new particle is invented. When axions remain undetected, another type of dark matter particle will be proposed. If supersymmetry doesn't appear in experiments, a new, even more complex theory will take its place. If dark matter continues to evade detection, new dimensions, forces, and hypothetical entities will be introduced instead.

This is not scientific progress—it is avoidance. Instead of asking whether more particles are truly necessary, physics assumes that the answer is always another missing piece. But what if particles are not the answer at all? What if physics is chasing ghosts because it refuses to consider a deeper energy-based framework? What if space is structured in a way that eliminates the need for extra particles altogether?

The reason physics never stops adding new candidates is clear: the Standard Model is not built on fundamental reality. It is built on a flawed assumption that requires infinite corrections to stay afloat.

The Flawed Logic of Infinite Discovery: How the Standard Model Turns Physics into a Particle-Hunting Game Instead of a Search for Fundamental Truth

Imagine trying to fix a broken equation by endlessly adding new variables to balance it. That is exactly how the Standard Model operates. Every time something doesn't fit, a new particle or force is proposed to make the numbers work. Each new "fix" then creates more contradictions, requiring even more adjustments.

This is not how real science should function. Newton's laws were simple because they accurately described motion. Einstein's relativity simplified gravity because it was built on fundamental principles. The Standard Model, however, becomes more complex with every discovery, adding more parameters, more assumptions, and more patches rather than clarifying the nature of reality.

A good theory should reduce unknowns, not multiply them indefinitely. Yet, the Standard Model thrives on endless unknowns because every mystery means more funding, more experiments, and more hypothetical discoveries. The endless particle hunt has become a cycle of self-preservation rather than a genuine quest for understanding.

The truth is, the Standard Model is not successful—it is just convenient. It provides enough statistical accuracy to maintain its dominance. It delivers just enough "discoveries" to keep funding flowing. It ensures physics remains trapped in an endless loop of theoretical speculation that never actually leads to a fundamental explanation.

The real question is not, "What new particles are missing?" but "Why do we need these particles at all?" And the answer is simple: we don't. The model is broken, and no amount of particle-hunting will ever fix it.

The Real Answer: Recognizing That Energy Structure Replaces the Need for Force Carriers

For over a century, physics has been obsessed with particles because it refuses to acknowledge the most fundamental truth of all: reality is not made of discrete particles but of structured energy. Forces do not require carriers—interactions emerge naturally from energy fields. Mass does not need the Higgs boson—it is a result of energy equilibrium in structured space. Gravity is not a separate force—it is an effect of energy distribution in space. The nucleus does not need a strong force—it is an energy-stable structure, not a collection of force-bound particles.

Everything physics is trying to "discover" has already been explained—just not in the language of particles. Instead of endlessly chasing axions, sterile neutrinos, and WIMPs, physics should be studying the energy properties of structured space. Instead of inventing force carriers, it should be asking whether forces are merely the natural result of energy interactions. Instead of endlessly modifying the Standard Model, it should be abandoning it altogether in favor of an energy-based understanding of reality.

The illusion of completeness has lasted long enough. The future of physics will not be found in endless particle searches. Real physics will begin only when we stop searching for missing particles and start understanding the fundamental structure of energy.

Key Takeaways

- The **Standard Model of Physics** is not a fundamental theory but a **collection of mathematical patches** designed to fit experimental data.
- Every time an inconsistency appeared, physicists **introduced new particles** (like the Higgs boson) rather than questioning the model itself.
- The model describes **forces as particle exchanges (gauge bosons)** without explaining **why these forces exist** in the first place.
- Instead of offering a **unified understanding of reality**, the Standard Model has grown into an **overcomplicated mess** with arbitrary constants and exceptions.
- **Dark matter, dark energy, and other unknowns** are introduced as placeholders instead of seeking a deeper explanation of energy structures.

What If...? (Speculative Exploration)

- What if the **Standard Model is just a temporary mathematical framework**, not a true reflection of physical reality?
- What if **force carriers (bosons) do not exist**, and instead, forces emerge from the **structuring of energy in space**?
- What if physics had focused on **uncovering fundamental principles** instead of **adding new particles** to patch flaws?
- What if gravity, electromagnetism, and nuclear interactions **are not separate forces**, but different manifestations of a single energy field?

Why Is the Standard Model So Overcomplicated?

- **Mathematical consistency was prioritized over physical intuition**, leading to an artificial set of rules rather than a natural understanding.
- Each time an experiment contradicted the theory, **a new particle was added** rather than reconsidering fundamental assumptions.
- The discovery of the Higgs boson was treated as a **confirmation of the model**, even though it raised **more questions than answers**.
- The Standard Model still **cannot explain 95% of the universe** (dark matter and dark energy), proving its fundamental limitations.

Challenges & Open Questions

- If particles do not carry forces, **what truly governs interactions between matter and energy?**
- Could a **universal energy field** replace the need for force-carrying particles?
- How can we develop a model that **unifies gravity, electromagnetism, and nuclear interactions** without contradictions?
- What experimental evidence could support **a structured energy-based alternative** to the Standard Model?

Suggested Readings & References

- **Critiques of the Standard Model** – Why many physicists believe it is an incomplete theory.
- **Tesla's Vision of Energy Structures** – A perspective that challenges particle-based force interactions.
- **Chavan's Grand Unified Theory of Physics** – A radical approach to replacing the Standard Model with an energy-based understanding.

6: The Gravity Illusion

Einstein's Final Patch

For centuries, gravity has remained the most elusive phenomenon in physics. Newton described it as an invisible force pulling objects toward each other. Einstein replaced this idea with the concept of space-time curvature, arguing that mass distorts the fabric of space, guiding objects along curved paths. Modern physics still struggles to reconcile gravity with quantum mechanics, unable to bridge the gap between Einstein's smooth continuum and the discrete nature of the quantum world.

In the previous chapters, we exposed the illusions of force-carrying particles, wave-particle duality, and the endless patchwork of the Standard Model. Now, we turn to the biggest illusion of all—gravity itself. Did Einstein truly explain gravity, or did he simply replace one mystery with another? Is gravity an actual effect of space-time curvature, or is it the result of structured energy fields? And if general relativity works so well on cosmic scales, why does it completely break down at quantum levels?

For over a century, physics has treated gravity as a fundamental force, but what if this assumption is wrong? What if gravity is not a force at all, but an emergent effect of structured energy fields? What if space-time curvature is not a physical reality, but just another mathematical fix—one that hides the deeper mechanism behind gravitational interactions?

This chapter dismantles the final major patch in modern physics—Einstein's reinvention of gravity. We examine how general relativity conflicted with quantum mechanics from the beginning, why black holes expose the limitations of Einstein's space-time model, and how gravity naturally emerges from energy gradients rather than warped geometry.

Physics never truly understood gravity—it only built models to describe its effects. Instead of questioning its core assumptions, it kept adding new layers of complexity, turning gravity into an ever-deepening paradox rather than a fundamental truth. Now, it's time to strip away the final illusion and reveal what gravity really is.

6.1 Einstein's Dilemma – Why Gravity Had to Be Reinvented

Gravity has always been one of the most puzzling forces in physics. It dictates the motion of planets, yet it appears to act instantly across vast distances. It holds galaxies together, yet it is the weakest of all known interactions. It bends space, yet no one can explain what space truly is.

By the late 19th century, Newton's law of universal gravitation had successfully predicted planetary motion, the behavior of falling objects, and celestial mechanics. Yet, despite its accuracy, it had one glaring flaw—it described how gravity worked but never explained why. How does the Sun "pull" on the Earth from 150 million kilometers away? How does an apple "know" to fall toward the ground? What connects objects across empty space?

Newton provided equations that worked, but they lacked a mechanism. He himself admitted this problem, famously stating that the idea of gravity acting across a vacuum without a medium was "so great an absurdity that I believe no man who has in philosophical matters a competent faculty of thinking can ever fall into it." Even he recognized that his equations were merely a mathematical tool, not a fundamental explanation.

Einstein was not content with this. He rejected the idea that gravity could be instantaneous and sought a deeper, more physical explanation. He wanted a model that eliminated the concept of an invisible force acting at a distance and replaced it with a tangible mechanism. His solution was general relativity—a theory that discarded Newton's force-based gravity and replaced it with curved space-time.

At first, this was seen as a revolutionary breakthrough. It explained planetary orbits without requiring an invisible pull. It correctly predicted the bending of light near massive objects. It became the foundation of modern cosmology. But despite its mathematical success, general relativity introduced a new set of contradictions—ones that physics has yet to resolve.

Newton's Gravity Worked—But It Had No Mechanism

Newton's law of gravity remains one of the most successful equations in science. It accurately describes the motion of planets and moons, explains why objects fall to the ground, and works so well that we still use it today to navigate spacecraft. Yet, there was one fundamental flaw—it lacked a physical explanation.

Newton's gravity assumed that objects attract each other, but it never answered why they do so. What medium allows gravity to act across empty space? How does gravity propagate? Newton himself never claimed to have an answer. His equations worked, but they left physics with an unsolved mystery. If gravity is not a force, then what is it? If there is no gravitational "pull," then why do objects move? If space is empty, how does the Sun affect the Earth?

Einstein set out to answer these questions, and in doing so, he rejected the idea of instantaneous gravitational interactions.

The "Spooky Action" of Instant Gravity—Why Einstein Rejected It

Newton's model suggested that gravity acted instantly, regardless of distance. If the Sun suddenly disappeared, Newton's equations implied that Earth would immediately feel the loss of gravity. There would be no delay, no transmission time—gravity would act faster than light. To Einstein, this was unacceptable.

Physics had already shown that light had a speed limit. Why would gravity be any different? How could two objects interact across empty space with no delay? What was the medium through which gravity "traveled"? To Einstein, Newton's gravity seemed to violate everything physics knew about how interactions should work.

He proposed an alternative: mass does not "pull" objects, but rather bends the fabric of space-time. Objects do not feel a force but instead follow curved paths in this distorted space. This eliminated the need for instantaneous action. Instead of gravity "pulling" Earth toward the Sun, Einstein suggested that Earth simply moved along the curve of space-time created by the Sun's mass.

This was the birth of general relativity, and at first, it seemed to solve everything.

How Einstein's Search for a Deeper Explanation Led to Space-Time Curvature

Einstein's new theory completely transformed our understanding of gravity. Instead of treating space as empty, he described it as a flexible fabric. Instead of an invisible force, he described gravity as the bending of this fabric. Instead of instantaneous action, he made gravity propagate at the speed of light.

The concept was simple:

- Massive objects distort space-time.
- Other objects move along these curved paths.
- This creates the effect we call gravity.

For example, the Sun bends space-time, and Earth follows a curved trajectory around it. A black hole bends space so much that even light cannot escape. Light passing near a massive object follows a curved path due to space-time distortion.

It was a brilliant mathematical framework. It correctly predicted the bending of starlight during a solar eclipse. It explained why Mercury's orbit deviated slightly from Newtonian predictions. It became the foundation for modern cosmology.

But despite its success, general relativity introduced a new set of contradictions. It was mathematically elegant, but physically incomplete.

Why General Relativity Was a Mathematical Success but a Physical Contradiction

Einstein's equations worked so well that physicists stopped questioning them. If the math works, why look for a deeper explanation? If the theory predicts experiments, why challenge the assumptions? If general relativity is so powerful, why consider alternatives?

This blind acceptance created a major problem. General relativity is mathematically perfect, yet it fundamentally contradicts quantum mechanics. It predicts singularities—points of infinite density that defy physics. It fails to explain gravity at small scales, leading to paradoxes.

This means that Einstein's gravity is not the final answer. Space-time curvature may describe the effects of gravity, but it does not explain its cause. Gravity is not about bending space—it is about structured energy. Einstein was right to reject Newton's "spooky action at a distance," but his own solution—curved space-time—was just another patch.

It did not explain gravity; it only replaced one mystery with another. It worked mathematically, but it ignored deeper physical realities. It solved one problem while creating new contradictions.

This means we must go deeper. If Newton's force was an illusion, and Einstein's space-time curvature is incomplete, then what is gravity?

Next: How Einstein's Gravity Contradicts Quantum Mechanics

- Why general relativity and quantum mechanics remain fundamentally incompatible.
- Why gravity does not behave like any other force in physics.
- How energy density—not space-time curvature—can finally explain gravity.

Because the truth is simple:

- Gravity was never a force.
- It was never space-time curvature.
- It was always an effect of energy structure.

6.2 How Einstein's Gravity Contradicts Quantum Mechanics

For over a century, physics has been built upon two towering theories: general relativity, which describes gravity and the structure of space-time at cosmic scales, and quantum mechanics, which governs the behavior of particles, forces, and energy at microscopic scales. Each works with astonishing precision within its own domain—relativity predicts planetary motion, black holes, and light bending near massive objects, while quantum mechanics explains atomic structure, electromagnetic interactions, and the fundamental properties of matter.

But despite their individual successes, these two theories remain fundamentally incompatible. General relativity treats space-time as a smooth, continuous fabric, while quantum mechanics insists that everything—including space itself—is quantized and fluctuating. Relativity describes gravity as curved space, while quantum mechanics suggests that all forces arise from discrete field interactions. The core assumptions of each theory directly contradict the other.

This creates a paradox: if general relativity is correct, then quantum mechanics must be incomplete. If quantum mechanics is correct, then general relativity must be wrong. Yet physics refuses to abandon either, choosing instead to pile on new mathematical fixes, additional particles, and ever more complex equations. But the failure to unify these two frameworks is not a failure of mathematics—it is a failure of fundamental assumptions.

The truth is simple: gravity is not a force, not a curvature, but an emergent effect of structured energy.

The Fundamental Contradiction: Why Relativity and Quantum Mechanics Cannot Be Unified

Einstein's general relativity was designed to describe gravity on large scales. It treats space as a continuous entity that bends in response to mass, assumes gravity acts smoothly across space without quantum fluctuations, and describes a universe where energy and matter curve space-time, causing objects to follow these curves. In contrast, quantum mechanics describes the universe at small scales, treating space itself as a fluctuating energy field, where particles and forces emerge from probabilistic interactions rather than smooth geometry.

The fundamental problem? Relativity requires space-time to be smooth, while quantum mechanics requires it to be discrete and dynamic. They cannot be merged into a single framework without breaking one of them. This is why, after decades of effort, no successful quantum theory of gravity has ever emerged. Every attempt to merge the two—string theory, loop quantum gravity, and others—has failed, because relativity and quantum mechanics are built on incompatible principles.

Why Gravity Does Not Behave Like the Other Fundamental Interactions

Modern physics categorizes four fundamental interactions: electromagnetism, the strong nuclear force, the weak nuclear force, and gravity. The first three are part of quantum mechanics, each mediated by force-carrying particles—photons for electromagnetism, gluons for the strong force, and W and Z bosons for the weak force. These forces all weaken in predictable ways over distance and can be expressed through quantum field interactions.

Gravity, however, behaves entirely differently. It has no confirmed force carrier, does not weaken in the same way as the other forces, and operates over infinite distances without a quantum mechanism. If gravity were truly a force, it should have a force-carrying particle (the hypothetical graviton), interact with the quantum vacuum like electromagnetism, and follow the same mathematical structure as other interactions. But none of this happens.

This is not a minor inconsistency—it is a fundamental breakdown of the Standard Model's framework. No graviton has ever been detected, and gravity does not behave like a quantized force. Einstein's model suggests that gravity is purely geometric, but this contradicts the way all other forces work in quantum mechanics. The reason is simple: gravity is not a force like electromagnetism. It does not need a force carrier because it does not operate through particle interactions—it emerges naturally from energy gradients in structured space.

The assumption that gravity must be a "force" like the others was one of the greatest errors in physics.

The Quantum Problem: Why General Relativity Ignores Energy Fluctuations in Space

Quantum mechanics tells us that empty space is never truly empty. Virtual particles constantly pop in and out of existence, fluctuations in energy density shift the quantum vacuum, and every force operates through interactions with quantum fields. But general relativity completely ignores this, treating space-time as a smooth, continuous fabric—even though quantum mechanics proves it is not.

Relativity assumes gravity is continuous, while every other force is quantized. It predicts singularities—regions of infinite density—where quantum mechanics suggests extreme energy fluctuations should prevent such infinities. If gravity were truly fundamental, it should interact with quantum fluctuations. A high-energy vacuum should produce gravitational effects, space should not be smooth but structured at quantum scales, and general relativity should account for quantum vacuum effects. But it does not.

Instead, relativity assumes space is empty except where mass is present. It cannot explain gravity at the atomic scale, treats it as a smooth field while every other interaction is discrete, and fails to reconcile with the known behavior of space in quantum mechanics. This means relativity is incomplete, and the idea of space-time curvature is just another patch—one that works only at large scales. If relativity cannot handle quantum effects, then it is not a fundamental theory. If gravity does not interact with the quantum vacuum, then space-time curvature is not the true cause of gravity. If relativity fails at small scales, then Einstein's model is only an approximation, not a deep explanation.

Why General Relativity Fails at Small Scales – The Breakdown of Einstein's Model

If general relativity were truly fundamental, it should work at all scales—from galaxies to atoms. But it completely fails at microscopic levels. Gravity is too weak to affect individual particles, has no measurable role inside an atom, and cannot explain why fundamental particles have mass.

For decades, physicists have tried to "quantize" gravity, attempting to force Einstein's theory to work at small scales. String theory was developed to merge quantum mechanics with relativity, but it has produced no verifiable predictions. Loop quantum gravity attempted to replace smooth space-time with discrete structures, but it still relies on relativity's assumptions. Every attempt to unify gravity with quantum mechanics has failed because relativity was never meant to be a quantum theory.

If a theory fails at small scales, then it is not fundamental. If gravity cannot be quantized, then space-time curvature is not the full answer. If relativity is incompatible with quantum mechanics, then something deeper must exist. The failure of Einstein's gravity at small scales is proof that it is just a mathematical approximation. It describes cosmic motion but does not explain the cause of gravity. It predicts large-scale effects but fails at quantum levels. It is a useful tool, but not a fundamental truth.

This means:

- Gravity is not space-time curvature.
- It is not a force like the other interactions.
- It is an effect of energy structure—just like every other fundamental interaction.

Next: The Failure of General Relativity Inside Black Holes

- Why Einstein's equations break down inside a black hole.
- How singularities prove that relativity is incomplete.
- Why black holes expose the limits of space-time curvature.

Because the truth is simple:

- Gravity is not a force.
- It is not curvature.
- It is an emergent effect of energy structure.

6.3 The Failure of General Relativity Inside Black Holes

Einstein's general relativity is often regarded as one of the greatest triumphs of physics. It redefined gravity, replacing Newton's mysterious "invisible force" with the curvature of space-time. It successfully explained planetary orbits, the bending of light, and even the expansion of the universe. It even predicted the existence of black holes—objects so dense that nothing, not even light, can escape their gravitational pull.

But inside black holes, Einstein's equations collapse. They predict singularities—points of infinite density and zero volume—where all known laws of physics cease to function. The equations generate mathematical infinities that have no physical meaning, offering no explanation for what actually happens inside a black hole. This is not a minor inconsistency—it is a fundamental failure.

A true theory of gravity should work everywhere, not just where it's convenient. A real model of space should not self-destruct into infinities. A fundamental law of nature should not break down inside the very objects it predicts. Black holes expose the limits of general relativity and prove that space-time curvature is not the full answer to gravity.

Why General Relativity Breaks Down Inside Black Holes

General relativity describes a black hole as a region where gravity becomes infinitely strong. The event horizon marks the point beyond which nothing can escape, while the singularity at the center is predicted to have infinite density and zero size. But at this point, general relativity collapses—its equations break down, space-time curvature becomes undefined, and physics no longer makes any sense.

If general relativity were truly fundamental, it should describe what happens at the singularity, explain how matter behaves inside the event horizon, and provide a physical mechanism for gravity in extreme conditions. Instead, it produces mathematical nonsense. A real object cannot have infinite density and zero volume—that is not physics, but a failure of the equations. Singularities contradict quantum mechanics, which demands a limit to energy density, and their appearance in relativity's equations is proof that the theory is incomplete.

Physics should not accept infinities as answers. Singularities are not real objects—they are signs that the theory itself is flawed. The fact that relativity breaks down inside black holes proves that it is just an approximation, not a fundamental truth. A complete theory of gravity must describe black hole interiors without producing infinities.

The Singularity Paradox: When Math Fails to Describe Reality

A singularity is not a discovery—it is a contradiction. It is what happens when a mathematical model fails to describe reality. General relativity predicts that inside a black hole, gravity becomes infinite, space-time collapses, and all physical laws break down. But physics tells us that no real object can have infinite density.

Whenever physics encounters infinities, it means one thing: the theory is wrong—or at least incomplete. If singularities were real, then the laws of physics would cease to function inside black holes. The very concept of space-time would collapse, and general relativity itself would become meaningless.

This tells us something profound: black holes expose a deeper layer of reality that goes beyond space-time curvature. If relativity cannot describe what happens inside a black hole, then it is not the final answer. There must be a deeper mechanism that explains gravity—one that does not rely on space-time bending infinitely.

Black Holes and the Limits of Space-Time Curvature

Einstein's gravity is built on the idea that mass bends space-time. The greater the mass, the stronger the curvature. At extreme densities, space-time bends infinitely, supposedly forming a singularity. But this logic collapses under its own weight.

If space-time can bend infinitely, why does it break down at singularities? If gravity is just curvature, why does modern cosmology require dark matter to explain galaxy motion? If space-time is smooth and continuous, why does quantum mechanics suggest that space itself is quantized?

The failure of general relativity inside black holes suggests something deeper: gravity is not just a geometric effect, and space-time is not a fundamental entity—it is an approximation. There must be an underlying energy structure that general relativity does not account for.

General relativity describes gravity, but it does not explain it. It predicts black holes but fails to describe what happens inside them. It produces equations that work—until they reach singularities. If a theory fails under extreme conditions, then it is not the complete answer. Just as Newton's gravity was replaced by Einstein's relativity, Einstein's relativity must now be replaced by a deeper understanding of gravity.

Why General Relativity's Failure in Black Holes Proves That Space-Time Is Not the Full Answer

The fact that relativity breaks down inside black holes is the strongest evidence that space-time curvature is not the fundamental cause of gravity. If singularities emerge from the equations, then general relativity is incomplete. If quantum mechanics cannot explain gravity, then something fundamental is missing. If space-time curvature fails inside black holes, then gravity must be something deeper than just geometry.

A true theory of gravity must work at all scales—from galaxies to black holes to quantum particles. It must explain mass, energy, and gravity without requiring infinite singularities. It must describe black hole interiors in a way that makes physical sense. This means moving beyond space-time curvature to find the real cause of gravity.

What if gravity is not a force, but an effect of energy structure? What if space is not an empty void, but a medium with measurable properties? What if singularities are not real, but just failures of our current models?

This is the next step in understanding gravity—not as the bending of space, but as a manifestation of structured energy.

Next: Why Gravity Is an Energy-Density Effect, Not Space-Time Curvature

- Why bending space-time is not enough to explain gravity.
- How structured energy fields naturally produce gravitational effects.
- Why space is not "nothing"—but an active, structured medium.
- How gravity emerges from energy gradients, not from force or curvature.

Because the truth is simple: Gravity was never about bending space—it was always about energy.

6.4 Why Gravity Is an Energy-Density Effect, Not Space-Time Curvature

For over a century, physics has treated gravity as a consequence of space-time curvature. Einstein's general relativity replaced Newton's force-based gravity with a geometric model, where the presence of mass bends space-time, guiding objects along curved paths. This elegant framework seemed to explain everything—planetary orbits, gravitational lensing, and even black holes.

But there is a deeper problem. Space-time curvature describes gravity, but it does not explain what causes it. General relativity ignores the quantum structure of space, and its failure in black holes and at small scales proves that something deeper must be at work. This means space-time curvature is not the true nature of gravity—it is only an approximation.

The real cause of gravity is not geometry, but structured energy. Gravity is not a force in the traditional sense, nor is it the bending of an abstract space-time fabric. It is an effect of energy density—a direct consequence of how energy is distributed in space. Instead of requiring a separate force, gravity emerges naturally from the structure of space itself.

Why Bending Space-Time Does Not Explain the True Nature of Gravity

Einstein's theory was a step forward, but it was not the final answer. Newton's gravity described attraction but had no underlying mechanism. Einstein's gravity described curved space-time but ignored the role of energy density. Both models predict motion but do not fully explain the cause of gravity itself.

The space-time model has serious flaws. It requires mass to bend space but never explains how mass interacts with space itself. It assumes space-time is smooth, while quantum mechanics suggests it is structured and fluctuating. It predicts singularities—places where space-time curvature becomes infinite—without offering a physical explanation.

If gravity were purely a result of space-time curvature, then why does quantum mechanics suggest that space itself fluctuates with energy? Why do black holes expose infinities that cannot be resolved by relativity? Why does general relativity fail to describe gravity at microscopic scales? These contradictions prove that gravity is not just about bending space—it must be something deeper, linked to the energy structure of space itself.

How Structured Energy Fields Naturally Produce the Effects We Call Gravity

One of the biggest mistakes in modern physics is treating space as "nothing." Space is not empty—it is filled with quantum fluctuations, energy fields, and virtual particles that constantly interact. Every known force operates through energy fields, yet gravity is still treated differently. If space has structure, then gravity must emerge from its energy distribution, not from an abstract bending of geometry.

Gravity is not a force acting at a distance, nor is it a warping of an empty void. It is an effect of energy density gradients in structured space. Mass does not bend space—it modifies the surrounding energy field. Objects move not because of an external force, but because they are following the natural pathways set by energy distribution.

This perspective resolves many of the issues that relativity struggles with. It explains why gravity is far weaker than other fundamental forces—it is not a discrete interaction but an emergent effect. It explains why no force carrier for gravity has been found—because gravity does not function like electromagnetism or the nuclear forces. It explains why general relativity breaks down in black holes—because it ignores the energy structure of space at extreme densities.

Gravity is not a mysterious bending of space-time. It is the natural result of mass interacting with the structured energy of space itself.

Why the Search for Gravitons Is Unnecessary—Gravity Emerges from Energy Gradients

For decades, physicists have searched for a graviton, the hypothetical force carrier of gravity. The Standard Model treats all other forces as being mediated by particles, so gravity is assumed to need one too. Various quantum gravity models attempt to define gravity as a particle-based force, with the graviton acting as the messenger. But the truth is simpler—gravitons do not exist because gravity is not a force like electromagnetism or nuclear interactions.

Electromagnetic forces require charge-based interactions, and thus, force carriers like photons. The strong and weak nuclear forces rely on particle exchanges to function. But gravity does not work this way—it is an emergent effect of energy density.

This means there is no need to quantize gravity in the way physics treats other forces. Instead of searching for gravitons, physics should be investigating the energy structure of space. If gravity emerges from structured energy, it does not need a separate force-carrying particle. The entire search for gravitons is based on a misunderstanding—assuming gravity must behave like the other forces, when in reality, it is something entirely different.

Just as space-time curvature was an illusion, so is the idea of a gravitational force carrier. Gravity is not transmitted by particles—it emerges from energy fields without requiring additional entities.

The Real Picture: Gravity as an Effect of Energy Structure, Not an Independent Force

Once we discard the illusion of space-time curvature and force carriers, gravity becomes far easier to understand. It is not a force—it is a response to energy gradients. It does not require particles—it is an effect of structured space. It does not bend space-time—it emerges from variations in energy density.

This perspective resolves the contradictions of modern physics. It explains why general relativity breaks down in black holes—because it ignores the quantum energy fluctuations that prevent infinite densities. It explains why gravity is fundamentally different from the other forces—because it is not a force at all. It explains why space is not empty, but an active, structured field that determines motion.

Mass does not curve space—it modifies the local energy field. Objects do not follow curved space—they move in response to structured energy distributions. Gravity is not a separate interaction—it is an emergent property of space's underlying structure.

This model bridges the gap between relativity and quantum mechanics without requiring force carriers, singularities, or extra dimensions. It unites gravity with quantum principles by treating space as an active, structured medium. It eliminates the need for gravitons, because gravity is not a force. It removes singularities, because energy structure prevents infinite densities from forming.

This is the real nature of gravity—not a force, not curvature, but an emergent effect of energy distribution.

Key Takeaways

- Einstein's **General Relativity** redefined gravity not as a force but as **the bending of space-time**, but this was a **mathematical construct, not a physical explanation.**
- The idea that mass warps space-time **explains motion but does not reveal the mechanism behind gravity itself.**
- Gravity is treated differently from other forces because it **does not fit into the Standard Model**, exposing the fundamental flaw in modern physics.
- The existence of **gravitational waves** is often seen as proof of relativity, but they do not explain **how mass actually interacts with space.**
- The failure to unify gravity with quantum mechanics suggests that **space-time bending may not be the real cause of gravity.**
- If gravity was truly a result of curved space-time, **why do we still need dark matter to explain missing mass?**

What If...? (Speculative Exploration)

- What if gravity is **not a force, nor a curvature of space**, but an effect of structured energy within the universe?
- What if **mass interacts with space through energy fields**, not by bending an abstract mathematical coordinate system?
- What if general relativity is just **a highly accurate approximation** rather than a fundamental truth?
- What if black holes are not **singularities in space-time**, but regions where energy structuring changes dramatically?

Why Did Physics Accept Gravity as Space-Time Curvature?

- **Newton's model of gravity** treated it as an invisible force, but Einstein **needed a new way to describe it mathematically**.
- The success of general relativity in predicting **planetary motion, light bending, and gravitational time dilation** made it difficult to challenge.
- **Curved space-time worked in equations**, so physicists accepted it even without a physical explanation.
- Einstein himself **was never fully satisfied** with his own theory, but physics moved forward with it because no better model existed.

Challenges & Open Questions

- If gravity is not curvature of space-time, **how does mass influence motion?**
- Could an **energy-based model** explain gravity more naturally?
- Why does general relativity fail **at quantum scales and in black holes?**
- If space-time is not fundamental, **what is the true nature of space?**

Suggested Readings & References

- **Einstein's Struggles with Gravity** – A historical look at Einstein's own doubts about general relativity.
- **Alternative Theories of Gravity** – Why modern physics still cannot unify gravity with quantum mechanics.
- **Chavan's Energy Based Relativity** – A new way to explain gravity without space-time curvature.

7: The Time Illusion

A Human Invention, Not a Fundamental Reality

For centuries, time has been treated as an absolute—a fundamental part of reality. We measure it, track it, and organize our entire existence around it. Physics treats it as a dimension, just like space, while relativity tells us time is flexible, yet we experience it as constant.

But what if everything we believe about time is an illusion? What if time is not a physical entity but a mental construct? What if the past, present, and future exist only in our perception, not in reality? What if time is simply an effect of motion, energy, and change—not something that exists on its own?

In the previous chapter, we dismantled the illusion of space-time curvature and revealed gravity as an emergent effect of structured energy. Now, we turn to an even deeper assumption—our belief in time itself.

Why does physics treat time as a fundamental dimension when it cannot be directly measured? Why does time disappear inside black holes if it is truly real? Why does quantum mechanics secretly assume an absolute time, even while relativity denies it?

This chapter will expose time not as a fundamental property of the universe, but as a human-created illusion. We will see how time emerges from energy interactions, not as an independent entity. We will explore how black holes prove time can vanish—meaning it cannot be fundamental. We will uncover the contradiction in quantum mechanics, where time is treated as absolute despite relativity's claims.

Time is not what we think it is. Physics has spent a century building equations around an illusion. Now, it's time to redefine reality—without time as a fundamental concept.

7.1 The Human Perception of Time – Why We Assume It Is Real

For most of human history, time has been treated as something real—something that flows, something that moves forward, something that we all experience. We wake up, go through our day, and assume time is passing. We age, objects decay, and clocks measure intervals, all reinforcing the belief that time exists as a fundamental entity. Science has built entire theories based on time as a fundamental dimension, using it as a framework for motion, causality, and even the structure of the universe.

But what if time is not an independent entity, but simply an illusion created by our perception? What if our brains are not sensing time but only tracking changes in the world? What if the past, present, and future are not real things but simply mental constructs? What if physics treats time as a dimension—not because it truly is one, but because it makes equations easier to solve?

How Our Brains Construct Time from Sensory Input

Time feels real because we experience change. We see objects move, hear sounds unfold in sequences, and feel our bodies aging. But what if we are not sensing time at all? What if our brains are only registering events happening in sequence? What if time is not something we "experience" but something we infer from memory?

Modern neuroscience suggests that our experience of time is an illusion created by the brain. The brain does not sense time directly—it only tracks changes in the environment. We perceive a "flow" of time because our memories allow us to compare past and present states. If our memory failed, we would have no sense of time at all.

Think about dreams. Inside a dream, a single moment can stretch into what feels like hours. Events can be out of order, yet still make sense in the dream. Time can speed up, slow down, or stop entirely. If time were truly fundamental, how could our experience of it be so flexible? Why does time feel different depending on our mental state? Why does it slow down during life-threatening moments but speed up when we are focused? The answer is simple: we are not sensing time itself—our brain is just interpreting change.

Time is not something flowing through reality. It is a mental construct—a way for our brain to make sense of motion and memory.

The Illusion of Past, Present, and Future – A Mental Construct, Not a Physical Reality

We assume the universe has a past, present, and future. The past is gone but recorded in memory, the present is where we exist, and the future has not happened yet but is expected. But this division exists only in our minds.

Physics has no "now"—the laws of nature work the same forward and backward. There is no universal present—what is now for one observer may be the past or future for another. Time is not flowing—only our perception makes it seem that way.

Consider a film reel. If you examine each frame individually, the "past" and "future" both exist at the same time. The illusion of motion happens only when the frames are viewed in sequence. What if the universe works the same way? What if all moments exist simultaneously, and time is just our way of experiencing them in order? What if the past is not "gone" but simply inaccessible from our perspective? What if the future is not "open" but already exists, waiting to be observed?

This idea aligns with the **block universe theory**—the concept that all of time exists at once, but we perceive it sequentially. Like watching a movie, we only see one scene at a time, but all the scenes already exist, whether we have seen them or not.

If this is true, then time does not "pass"—it is simply an effect of how we process reality. The future is already "there"—but we cannot access it until we move forward in our perception. The past has not disappeared—it is just no longer visible to us. The idea that time flows like a river is just an illusion of human experience.

Why Physics Treats Time as a Dimension, But Human Experience Suggests Otherwise

Physics treats time as a fourth dimension, just like space. In relativity, space and time are combined into space-time, where objects move through this four-dimensional structure. Time dilation shows that time is not constant—it can stretch and contract depending on motion and gravity.

But our experience of time is nothing like a dimension of space. We can move freely in space, but we cannot move freely in time. We can control our motion in space, but time moves forward no matter what we do. We can travel to different locations in space, but we cannot revisit the past.

If time were truly a dimension, why do we experience it differently from space? Why can we move forward and backward in space, but only forward in time? Why do objects age and decay but never reverse their states? Why is time relative for moving observers, while space remains consistent?

The reason is simple: time is not a dimension like space—it is an emergent effect of energy interactions. Physics treats time as a mathematical tool, not a physical entity. It is useful for equations, allows predictions and measurements, but does not exist as something independent.

The Paradox: If Time Is Real, Why Can't We Detect or Measure It Directly?

If something is real, we should be able to observe it, detect it, and measure it directly. We can see and measure space. We can measure mass, energy, and forces. We can detect electromagnetic waves, gravity, and subatomic particles. But what about time?

We can measure clocks, but clocks only track motion—not time itself. We can define time intervals, but these are based on vibrations of atoms—not an actual substance called "time." We can calculate time's effects, but we cannot isolate time as an independent entity.

If time were truly real, physics should be able to find a particle or field associated with time, detect a fundamental property of the universe that produces time, and explain what time is made of and how it exists. But none of this happens.

Time is always tied to motion, energy, and mass—but never detected by itself. The only way we measure time is by tracking how objects move or change. This means time is not a thing—it is an effect. Time is not fundamental—it is a concept we created to track change.

And if time does not exist as a fundamental reality, then physics must abandon the idea that time is a dimension. We must rethink the laws of motion and causality. We must replace time with a new framework—one based entirely on energy interactions.

Next: How Time Emerges from Mass-Energy Interactions

Now that we have seen how time is a human-made illusion, we must ask: what is actually happening when we think time is passing? If time does not flow, what causes the illusion of forward motion? If the past and future exist simultaneously, why can we only experience one moment at a time?

In the next section, we will explore how time emerges from mass-energy interactions. We will see why time does not flow, but is simply a product of motion. We will uncover how entropy creates the illusion of a one-way time direction. We will examine why energy determines the passage of time—not the other way around.

Because the truth is simple: **time is not real—it is a human-made illusion.**

7.2 How Time Emerges from Mass-Energy Interactions

For centuries, time has been treated as a fundamental entity—a flowing, independent dimension that governs reality. Einstein's relativity incorporates time as part of space-time, stretching and contracting under the influence of gravity. Our daily experience reinforces this view, making us believe that time moves forward, dividing past, present, and future. Clocks, calendars, and equations measure time, giving the impression that it is real.

But what if time is not something that exists on its own? What if time is not a fundamental dimension, but simply a result of how matter and energy interact? What if the passage of time is an illusion created by entropy—the natural tendency of systems to evolve? What if mass and energy determine how we experience time, instead of time determining how mass and energy behave?

This section dismantles the illusion of time as an independent entity and reveals its true nature—an emergent effect of energy and mass interactions.

Why Time Is Not an Independent Dimension—It Arises from Motion and Change

Physics treats time as a coordinate, just like space. In relativity, time is simply another dimension, linked to space in four-dimensional space-time. It can be stretched or compressed by gravity and motion, and objects move through time just as they move through space. But this assumption is flawed.

Unlike space, which exists independently, time only exists when something changes. Without motion, there is no way to measure time. If time were truly fundamental, it should exist independently of all motion and interactions. It should be measurable even in a static universe. It should be observable as a property of reality, just like mass or energy.

But in a motionless, energy-less universe, time ceases to exist. If there is no motion, there is no before and after. If there is no change, time has no meaning. If nothing happens, time is indistinguishable from non-existence. This proves time does not exist on its own—it only emerges when things interact, move, and evolve. Time is not a background framework—it is an effect of energy dynamics. It is not something we move through—it is something we infer from changes in the universe.

How Entropy Creates the Illusion of a One-Way Time Flow

One of the biggest mysteries in physics is why time only moves forward. We remember the past, but not the future. Everything ages and decays, but never reverses. Cause and effect always follow the same order. But why?

The answer lies in entropy—the natural tendency of systems to move toward disorder. In a closed system, things always progress from order to disorder. A hot cup of coffee cools down—it never spontaneously reheats. A shattered glass never reassembles itself. This creates the illusion of time moving forward.

We define the past as the state of lower entropy and the future as the state of higher entropy. We experience time as flowing forward because disorder always increases. But entropy is not time—it is just an effect of energy distribution. The universe does not "move forward in time"—it simply follows energy gradients. Time does not push events forward—events unfold because energy flows from high to low states.

This means time does not exist independently—it is just a measurement of entropy change. The so-called "arrow of time" is not a fundamental property of the universe but a result of energy dissipation. If entropy could be reversed, time would appear to reverse as well—but time itself is not a real entity.

Why Mass and Energy Define the Passage of Time, Not the Other Way Around

In relativity, gravity affects time. A clock runs slower in a strong gravitational field. Time stretches near black holes. Fast-moving objects experience time more slowly than stationary ones. Physicists interpret this as proof that time is fundamental, arguing that time is "real" because it can be stretched and compressed. Time dilation, they say, proves that time is a dimension, just like space.

But what if we are looking at it backward? What if mass and energy do not bend time, but instead determine how we measure it? What if the speed of atomic processes changes under different energy conditions, creating the illusion of time distortion? What if clocks slow down not because time itself is changing, but because energy interactions are changing?

If time were truly fundamental, nothing should affect it. But if time is just an effect of energy interactions, then it makes perfect sense that mass and gravity can alter our measurement of it. This means time is not controlling the universe—the universe is controlling our experience of time.

This explains why time slows down near black holes—because extreme gravity alters energy structures. It explains why time moves differently for high-speed objects—because motion affects energy states. It explains why there is no universal clock—because time is just a local effect of mass-energy interactions. Time does not control physics—physics controls what we perceive as time.

The Deeper Truth: Time Does Not "Flow"—It Is an Effect of Energy Dynamics

If time were real, it should exist independently of everything else. But it does not—remove energy and motion, and time disappears. It should be measurable as a fundamental entity—but it never is. It should behave like space—but it does not allow free movement.

Instead, time is an emergent property of energy interactions. It is a measurement of motion and change. It is a side effect of entropy and energy distribution. It is not something flowing—it is something we experience because of how energy operates.

This is why time "dilates" under gravity—because gravity alters energy, not time itself. It is why time stops in black holes—because energy structures collapse, not because time disappears. The universe does not "move through time"—it simply evolves as energy redistributes.

If time were fundamental, it would exist on its own, independent of everything else. But time is not fundamental—it is emergent. Time does not control events—events define what we call time. Physics does not need time—it needs energy-based models that describe reality without relying on a fictional dimension.

Next: Why Time Disappears in Black Holes, Proving It Is Not Fundamental

Now that we understand time is an emergent effect of energy, we must ask: what happens when time vanishes? If time were fundamental, black holes should not be able to "stop" it. If time were an independent entity, it should continue even in extreme gravitational fields. But black holes reveal something different.

In the next section, we will explore why time "stops" at the event horizon—but energy remains. We will uncover how black holes expose the illusion of time as a real entity. We will examine what happens inside a black hole when time is no longer a factor.

Because the truth is simple: **time is not a thing—it is an illusion of energy motion.**

7.3 Why Time Disappears in Black Holes, Proving It Is Not Fundamental

For all of human history, we have assumed that time is a constant presence—something that flows through the universe, governing every event. We measure it with clocks, track it in equations, and use it to define cause and effect. Physics treats it as a dimension, woven into the fabric of space-time. We assume it is always there—everywhere, for everything.

But black holes challenge this assumption. General relativity predicts that time slows down near a black hole. At the event horizon, time appears to stop completely for an external observer. Inside the black hole, time loses all meaning—physics can no longer describe what happens.

This leads to a fundamental paradox. If time were truly real, it should not be able to "stop." If time were fundamental, it should exist inside a black hole just as it does outside. If time is required for cause and effect, then how can events still occur inside a black hole after time disappears?

Black holes do not just bend space—they expose the illusion of time itself. If time can vanish, then time is not fundamental. If physics breaks down without time, then physics has been built on a flawed assumption. If we want to understand the true nature of the universe, we must stop relying on time as a primary concept.

General Relativity Predicts That Time Stops at the Event Horizon

According to Einstein's equations, time slows down in strong gravitational fields. If you are near a black hole, time moves slower for you than for an observer far away. At the event horizon—the point of no return—time slows to a complete stop from an external perspective. This means that, from the viewpoint of an outside observer, nothing ever actually crosses the event horizon—it just "freezes" there.

But what happens inside? If time stops, how can anything still exist or happen? How can an object fall into a black hole if time no longer progresses? If time ceases to move, does the black hole contain a frozen record of everything that ever fell in?

These are not just theoretical questions—they reveal a deep flaw in our understanding of time. Physics cannot describe what happens beyond the event horizon. Einstein's equations stop making sense—time becomes undefined. If a fundamental law breaks down in extreme conditions, then it is not truly fundamental.

This is not just a problem for relativity—it is a problem for the entire foundation of physics. If time were real, it should continue to function in all conditions. If it can stop or disappear, then it was never a fundamental property of the universe. Black holes reveal that time is just a mathematical tool—not an actual, existing thing.

The Paradox: If Time Can Vanish, How Can It Be a Fundamental Property of the Universe?

A fundamental property of the universe should not be able to vanish. Mass does not "stop" in strong gravity—it remains measurable. Energy does not "stop" in extreme conditions—it continues to interact. Space does not "stop"—it still exists beyond the event horizon.

But time does stop. It reaches zero at the event horizon, making further calculations impossible. It loses meaning inside the black hole—there is no "before" or "after." It becomes irrelevant—meaning the laws of physics, which depend on time, also collapse.

This is not what happens to a real, physical entity. If time were real, it should behave like space, mass, or energy. But instead, it behaves like an effect—something that disappears under extreme conditions. This proves that time is not a fundamental part of reality—it is only a derived concept.

If time can vanish, then it is not required for existence. It is not necessary for physical interactions. It is not an independent part of the universe—it is just a way we describe energy interactions. This means that all of physics' reliance on time must be reexamined.

If time is not fundamental, then what governs motion and change? If time is not real, what actually happens inside black holes? If time does not flow, how do events still unfold? The answer lies in the true nature of energy, structure, and interaction.

How Black Holes Expose the Limits of Time-Dependent Physics

All of modern physics depends on time. Newtonian mechanics requires time to define motion. Relativity treats time as part of space-time. Quantum mechanics assumes a fixed background time for equations to work.

But inside a black hole, Newtonian time is meaningless—objects can no longer move in the normal sense. Relativistic time stops—meaning relativity cannot describe what happens. Quantum equations become useless—since they rely on a background time that no longer exists.

If a theory collapses in extreme conditions, it is incomplete. A true model of reality should work everywhere. Physics should not break down inside black holes—it should explain them. If time disappears in a black hole, then any theory built on time is flawed.

Black holes prove that our entire framework of physics must change. Time-dependent equations are temporary fixes, not fundamental truths. Space-time is just a mathematical tool, not a real physical entity. Physics must move beyond time—and find a deeper explanation.

The Missing Piece: If Time Does Not Exist Inside a Black Hole, What Replaces It?

If time is an illusion, then something else must govern reality. Objects inside a black hole still exist—even though time has stopped. Mass and energy remain—even though time is no longer defined. Interactions continue—suggesting that something deeper controls motion and change.

The answer is energy structure. If time is not fundamental, then the evolution of the universe must be driven by energy dynamics. If time is just an effect of entropy, then we need a new framework based on energy flows. If physics collapses without time, then we must replace time-dependent models with energy-based models.

This is the next step in redefining reality. Gravity is not a force—it is an effect of energy structure. Time is not a dimension—it is an illusion created by motion and entropy. The universe does not "move through time"—it simply evolves according to energy distributions.

Inside a black hole, space and time collapse—but energy remains. This proves that energy, not time, is the fundamental driver of reality. The "flow" of time is just an effect of energy redistribution. If we remove time from physics and replace it with energy-based interactions, the universe makes far more sense.

This means that physics must abandon time as a fundamental entity. Space-time was an approximation—not a deep truth. Time-dependent equations are mathematical conveniences—not physical laws. Reality must be redefined—not with time, but with structured energy as the fundamental framework.

Next: How Quantum Mechanics Secretly Treats Time as Absolute, Contradicting Relativity

Now that we have seen how black holes expose the illusion of time, we must look at the contradiction in quantum mechanics. If relativity denies absolute time, then why does quantum physics still rely on it?

In the next section, we will explore why quantum mechanics assumes a fixed time, even though relativity denies it. We will uncover how physics contradicts itself by treating time differently in different theories. We will examine why the failure to quantize time proves it is not a real entity.

Because the truth is simple: **Time does not exist—it is just an effect of energy motion.**

7.4 How Quantum Mechanics Secretly Treats Time as Absolute, Contradicting Relativity

For over a century, physics has struggled to reconcile two of its greatest theories—general relativity, which describes gravity and large-scale cosmic structures, and quantum mechanics, which governs the behavior of particles at microscopic scales. Both theories work incredibly well in their own domains. Relativity tells us that time is flexible, stretching and contracting depending on gravity and motion. Quantum mechanics, on the other hand, treats time as absolute—a fixed background against which all particles evolve.

This is one of the biggest contradictions in modern physics. If time is relative in one theory but absolute in another, which one is correct? If time can bend and stretch in relativity, why does quantum mechanics ignore these effects? If quantum mechanics does not depend on relativity's version of time, then is time even real?

This section reveals how quantum mechanics secretly treats time as an absolute parameter—even though relativity proves it is not. It exposes how quantum physics depends on an external "clock" that is never explained. It shows how the failure to quantize time proves that time is not a fundamental entity. It asks the deeper question: if time were real, why do the most fundamental quantum laws completely ignore it?

The Strange Contradiction: Relativity Says Time Is Relative, But Quantum Mechanics Assumes a Fixed Time

Relativity and quantum mechanics do not agree on the nature of time. In relativity, time is flexible—it stretches in strong gravity and slows down at high speeds. In quantum mechanics, time is a rigid parameter—it does not change, bend, or respond to gravity.

This contradiction is rarely discussed, but it is one of the biggest flaws in modern physics. If time is truly relative, quantum mechanics should account for time dilation—but it does not. If time is truly absolute, relativity should not show time slowing down under gravity—but it does. This means one of these theories is built on a false assumption.

Physics has never resolved this contradiction. Instead of fixing the problem, physicists simply treat time differently depending on which theory they are using. This inconsistency is not a minor issue—it is a fundamental flaw. If time were truly fundamental, it should behave the same way in all of physics—but it does not.

This is the clearest evidence that time is not real. Relativity treats time as a changing property of space-time. Quantum mechanics treats time as an external constant. If two of our most powerful theories cannot agree on time, then time is not fundamental—it is an illusion.

Why Quantum Equations Rely on an External "Clock" That Is Never Explained

In quantum mechanics, time is not an active player. It is not an operator like position or momentum. It is not quantized—it is treated as a smooth, unchanging background. Quantum equations work only if time is assumed to be absolute and fixed.

But where does this time come from? What defines the "ticking" of time in quantum mechanics? If time is relative in relativity, why doesn't quantum physics account for time dilation? If the universe has no absolute clock, why do quantum equations assume one?

The truth is simple. Quantum mechanics requires an external clock—but it never defines what that clock is. Every quantum system evolves in time, but the time parameter itself has no physical meaning. If time were truly real, quantum physics should explain its nature—but it does not.

This is a massive oversight. If time is fundamental, quantum mechanics should have a mechanism to define it. If time is relative, quantum mechanics should incorporate time dilation. Instead, quantum mechanics assumes time is an unchanging background—directly contradicting relativity.

This is not how fundamental concepts behave. Space, mass, and energy all have physical definitions. Forces like electromagnetism and gravity have measurable effects. But time has no clear definition, no cause, and no physical existence.

This suggests time is not a real thing—it is just a convenient placeholder in equations. Quantum physics does not explain time—it ignores it. This is not a sign of a deep truth—it is a sign of a broken assumption.

How the Failure to Quantize Time Exposes Its Non-Fundamental Nature

One of the biggest goals of modern physics is to merge quantum mechanics with gravity. All other forces in physics—electromagnetism, strong nuclear, weak nuclear—are quantized. If gravity is to be unified with quantum mechanics, time should also be quantized. But every attempt to quantize time has failed.

Why? Because time is not real—it is just an effect of energy interactions. There is nothing to "quantize" because time is not a physical entity. Physicists have been trying to force a solution to a problem that does not actually exist.

The failure to quantize time proves that time is not a fundamental dimension. Time does not behave like other physical quantities. Time is just a mathematical tool—not a real part of the universe.

If time were truly real, there should be a smallest unit of time, like the Planck length is for space. Quantum mechanics should be able to describe how time emerges from energy. But no experiment has ever found a fundamental "quantum of time."

This means physics must abandon time as a fundamental concept. Time is not something that "flows"—it is an effect of how energy interacts. The fact that time cannot be quantized proves it is not a fundamental part of reality.

The Real Question: If Time Were Real, Why Do the Deepest Quantum Laws Ignore It?

If time is fundamental, why does quantum mechanics treat it as a fixed background? Every quantum equation depends on time—but time itself has no quantum properties. Particles and forces are quantized—but time is not. Quantum mechanics requires time to exist—but it does not define what time actually is.

This is the greatest contradiction in physics. A fundamental property should not be ignored in the deepest laws of the universe. If time were real, it should be treated the same way as space, mass, and energy. Instead, time is just a parameter that physicists plug into equations without explanation.

This proves that time is not a real dimension—it is just a measurement of energy changes. Quantum mechanics does not explain time because time does not need to be explained. The reason time cannot be defined is simple—because it does not exist as a fundamental entity.

Physics must now move beyond time. Time was useful for equations, but it was never real. The next era of physics must replace time with a deeper framework—one based on energy interactions. We must stop treating time as an independent reality and start seeing it for what it is—an illusion created by motion and energy.

Key Takeaways

- **Time is not a physical entity** but a measurement tool humans created to describe change.
- Physics treats time as a **real dimension** in space-time, yet there is no evidence that time has a physical existence.
- **Relativity distorts time** based on speed and gravity, but this does not mean time itself is a substance—it only reflects changes in energy interactions.
- **The past and future do not physically exist**—only the present moment is real.
- The idea of **time flowing** is a human perception; physics has no evidence of an absolute "flow" of time.
- If time were truly fundamental, **it should exist independently**, yet it can only be measured through motion and energy exchanges.

What If...? (Speculative Exploration)

- What if **time is just an effect of change**, not a separate dimension?
- What if relativity's time dilation is **not a change in time itself** but a shift in energy interactions?
- What if removing time from physics equations leads to **a more accurate understanding of reality**?
- What if humans evolved to perceive time **only as a survival mechanism**, rather than it being a fundamental property of the universe?

Why Did Physics Treat Time as a Real Entity?

- Newton assumed time was **absolute and universal**, making it a key part of classical mechanics.
- Einstein introduced **time as a fourth dimension**, but this was a mathematical convenience, not a proof of time's physical existence.

- Modern physics struggles with time because **quantum mechanics and relativity treat it differently**.
- **The idea of a "block universe"** (where past, present, and future exist simultaneously) was accepted because it made relativity work mathematically.
- People experience the passing of time, but **experience does not prove time is fundamental—only that we observe change**.

Challenges & Open Questions

- If time does not exist as a fundamental entity, **how do we explain time dilation in relativity?**
- Could physics be rewritten using **only energy interactions** without the need for time?
- What is the true nature of **causality** if time is not fundamental?
- If time does not exist, **what happens to concepts like entropy and the arrow of time?**

Suggested Readings & References

- **Einstein's Relativity and the Nature of Time** – A historical analysis of how time became a dimension in physics.
- **The Illusion of Time in Quantum Mechanics** – Why some physicists believe time is not real.
- **Chavan's The Blueprint of the Universe** – A deeper explanation of reality without treating time as fundamental.

8: Einstein vs. Tesla

The Suppressed Debate on Energy and Space

For over a century, Albert Einstein has been celebrated as the genius who reshaped our understanding of the universe, while Nikola Tesla has been remembered as a brilliant yet eccentric inventor, a visionary whose ideas were ahead of his time but largely dismissed by mainstream science. Einstein's theory of relativity redefined space, time, and gravity, becoming the foundation of modern physics. Meanwhile, Tesla's vision of an energy-filled universe—one where space itself was alive with structured forces—was rejected as speculative and unscientific.

Einstein became the face of theoretical physics, his equations dominating the way we understand reality. Tesla, on the other hand, faded into obscurity, his work overshadowed by the rise of relativity and quantum mechanics. But what if history unfolded differently? What if Tesla was right in rejecting Einstein's space-time model? What if Einstein himself, despite championing relativity, harbored doubts about his own framework and suspected that Tesla's structured vacuum energy theory had merit—but could never openly admit it?

Modern physics made a critical choice—one that may have been a mistake. It embraced Einstein's elegant but abstract mathematical descriptions and ignored Tesla's insights into energy and the structure of space. By choosing geometry over energy, it set physics on a path that, a century later, remains filled with contradictions.

In the previous chapters, we dismantled the illusions of time, space-time curvature, and the flawed patches of quantum mechanics. Now, we turn to one of the greatest intellectual battles in history—one that was never openly fought, yet silently shaped the future of physics. It was Tesla versus Einstein, a conflict between two radically different views of the universe: one based on dynamic energy, the other on abstract geometry. It was a debate that could have rewritten physics, but was never allowed to unfold.

This chapter uncovers the suppressed debate between two of the greatest minds in history. We will explore how Tesla rejected Einstein's relativity and instead supported a structured vacuum filled with energy. We will question whether Einstein, despite his public stance, secretly entertained doubts about space-time and whether he recognized Tesla's ideas as valid but impossible to integrate into the physics of his time. Most importantly, we will examine what modern physics lost by dismissing Tesla's energy-based theories and why rediscovering them could be the key to fixing science today.

History has told us that Einstein was right and Tesla was wrong. But what if history itself was wrong?

8.1 The Forgotten Rivalry – Why Tesla and Einstein Were Opposites in Thought

Science has always been a battlefield of competing ideas. Some theories survive and define the future, while others are buried, forgotten, or deliberately suppressed. Einstein's theory of relativity became the cornerstone of modern physics, shaping our understanding of space, time, and gravity. Tesla, on the other hand, envisioned a universe filled with structured energy—a vast, interconnected field governing all motion and interactions. His ideas were dismissed as speculative, and his work was pushed aside. One man became the face of theoretical physics, while the other was reduced to a brilliant but misunderstood inventor.

Yet, beneath this historical narrative lies an unresolved debate—one that was never openly fought, but that could have changed the course of physics entirely. Einstein and Tesla saw the universe through completely different lenses. One built his theories on space-time geometry, while the other believed energy was the fundamental fabric of reality. One rejected the idea of an energy medium, while the other insisted that space was anything but empty. Their views were not just different—they were irreconcilable. If one was right, the other had to be wrong.

This is the story of two opposing worldviews—one that shaped modern science and one that was ignored.

How Einstein and Tesla Viewed the Universe Through Completely Different Lenses

At first glance, Einstein and Tesla seemed to have little in common. Einstein was a theoretical physicist, a master of mathematical abstraction. Tesla was an experimentalist, obsessed with tangible results and real-world applications. Einstein built elegant equations to describe the universe, while Tesla built machines to harness its forces. But their greatest difference lay in how they understood space itself.

Einstein saw space as a mathematical construct—empty, but flexible, bending under the influence of gravity. Tesla, however, believed space was a structured medium, filled with energy that governed all motion and interactions. This was not a minor disagreement—it was a complete contradiction. If Einstein was right, then Tesla's structured energy-filled universe could not exist. If Tesla was right, then Einstein's relativity was just an elegant illusion.

Both men changed the world with their ideas—but only one vision survived.

Why Tesla Rejected Space-Time and Believed in an Energy-Filled Vacuum

Tesla never accepted Einstein's theory of relativity. He dismissed the idea of space-time curvature as unnecessary mathematical complexity, arguing that it explained motion but failed to explain the true nature of space. He believed that the universe was governed by structured energy fields, not by the bending of an abstract space-time fabric.

His reasoning was simple. If space were truly empty, how could electromagnetic waves travel through it? If space had no structure, how could energy move across vast distances? If gravity were nothing more than curved space-time, why did it behave like a force with measurable energy effects?

Tesla insisted that space was not a void but a dynamic, structured medium. He called it the "luminiferous ether"—a sea of energy filling the cosmos. He believed this energy could be harnessed, transmitted, and manipulated, eliminating the need for traditional fuel-based power. His experiments with wireless electricity suggested that energy could move freely through space itself, defying the conventional understanding of physics. He even believed it was possible to extract limitless energy from the vacuum—a concept that modern physics is only now beginning to reconsider.

Yet, instead of being explored, Tesla's ideas were rejected. Einstein's relativity eliminated the need for an energy-filled space, and the failure of the Michelson-Morley experiment to detect the ether was used as proof against Tesla's structured vacuum. Physics moved toward space-time, leaving Tesla's energy-based model behind.

But what if this was the greatest mistake in scientific history?

How Einstein's Relativity Rejected the Need for a Medium, While Tesla Insisted Space Had Structure

Einstein's theory of relativity did not initially aim to reject Tesla's energy-based vision—but it ultimately did. Early in his career, Einstein himself searched for a medium that could explain light and gravity. But when the ether was "disproven," he removed the need for it altogether. Instead of an energy-filled vacuum, he proposed space-time—a four-dimensional fabric that could stretch and bend under the influence of mass.

This shift was profound. It turned space into a purely mathematical concept, detached from any physical substance. It redefined gravity as geometry rather than an energy effect. It abandoned the idea of a structured vacuum and replaced it with an abstract void.

Einstein's equations worked beautifully—but at a cost. They could describe motion, but they could not explain what space truly was. They treated light as both a wave and a particle, yet ignored the need for a wave medium. They mathematically modeled gravity but left it disconnected from all other forces in physics.

Tesla saw this as a dangerous step in the wrong direction. He argued that space could not be empty—it had to contain structured energy. He believed that motion, force, and electromagnetism all required an underlying medium to function. To him, relativity was not just incorrect—it was misleading physics away from reality.

He warned that if physics abandoned the study of structured energy, it would become a purely mathematical exercise, detached from practical discoveries. He feared that by removing energy as the foundation of reality, science would blind itself to fundamental truths about how the universe actually works.

Tesla was right. A century later, modern physics still has no clear definition of what space is. Relativity describes space-time but does not explain its substance. Quantum mechanics talks about energy fields but refuses to call space a structured medium. The result? Physics remains fragmented, unable to unify its theories into a coherent picture of reality.

The Suppressed Question: What If Tesla Was Right About Energy, and Einstein Was Forced to "Fix" Physics?

History tells us Einstein was right and Tesla was wrong. But history is written by the winners. And in science, the "winner" is often the theory that fits the mathematical models best—not the one that explains reality most clearly.

What if Tesla's structured vacuum energy was the missing piece all along? What if Einstein knew relativity was incomplete but could not challenge the system he had built? What if physics abandoned Tesla's insights too soon, locking itself into a dead-end model?

There is evidence that Einstein himself had doubts about relativity. Later in life, he struggled with the concept of the quantum vacuum—a strange sea of fluctuating energy that bore a striking resemblance to Tesla's structured space. He never found a way to unify gravity with quantum mechanics—perhaps because space-time and energy fields do not fit together. He may have realized too late that Tesla's vision of an energy-based universe was not just plausible—but necessary.

Today, physics remains stuck. Relativity and quantum mechanics contradict each other. Gravity remains the only force without a clear quantum description. Patches like dark matter and dark energy have been introduced to fix the failures of the space-time model. But what if the answer was always there?

What if Tesla's energy-based vision was suppressed, and we are only now beginning to rediscover it? What if space is not a void but a structured, energetic medium? What if relativity was just a temporary fix, and physics must return to Tesla's unfinished work?

If so, then everything must be reexamined—from gravity to electromagnetism to the very fabric of reality itself. Tesla was not wrong—he was ignored. And the future of physics may not lie in further refining space-time but in rediscovering the structured energy that Tesla saw all along.

8.2 How Tesla Rejected Relativity and Supported Structured Vacuum Energy

Nikola Tesla was not just an inventor—he was a visionary scientist who saw the universe through an entirely different lens than mainstream physics. Unlike Einstein, who redefined space as a mathematical structure, Tesla believed that space was a vast ocean of structured energy, waiting to be harnessed. He did not merely speculate about this—he built experiments and devices that directly challenged the foundations of modern physics. Yet, despite his brilliance, Tesla's insights into the nature of space and energy were ignored, dismissed, and ultimately erased from the mainstream scientific narrative.

Why did Tesla believe Einstein's relativity was wrong? What did he discover about space that Einstein's equations could not explain? Why does modern physics still refuse to acknowledge Tesla's energy-based model of the universe? This section uncovers the missing physics behind Tesla's work and explores why his rejection of relativity may have been justified.

Why Tesla Called Einstein's Relativity a Mathematical Illusion

Tesla was never one to soften his opinions. He openly dismissed Einstein's relativity as an unnecessary abstraction that had misled physics. In his view, relativity did not explain the workings of the universe—it merely complicated them with elegant but impractical mathematics. Tesla rejected the notion that mass could bend space and argued that gravity and electromagnetism shared a common energy-based origin, which Einstein had completely ignored.

He famously criticized relativity, stating, *"Einstein's theories are magnificent mathematical garb which fascinates, dazzles, and makes people blind to the underlying errors."* He went even further, calling relativity *"a beggar wrapped in purple whom ignorant people take for a king."* To Tesla, Einstein's theory was not a profound discovery, but a mathematical trick that obscured the true nature of space and energy.

Tesla's rejection of relativity was not without reason. He believed that relativity had removed a critical component of physics—the concept of a structured medium in space. Instead of viewing gravity as an effect of bent space-time, Tesla saw it as a direct result of energy interactions within a structured vacuum. To him, physics had taken a wrong turn by embracing geometry over real, physical forces.

Tesla was not alone in questioning relativity. Even Einstein himself had lingering doubts, spending years attempting to modify his own theory. Many early physicists struggled with relativity's inability to connect with quantum mechanics. But while others remained silent or cautious, Tesla saw the problem early and was not afraid to challenge the mainstream narrative.

Tesla's Belief That Space Was Not Empty, But an Ocean of Energy Waiting to Be Tapped

Tesla's view of the universe was fundamentally different from Einstein's. While Einstein described space as a void that could bend under mass and gravity, Tesla saw it as an active, energy-filled medium. To Tesla, energy was not something that had to be generated—it was already present everywhere, embedded in the very fabric of space itself.

He argued that space was not empty, but full of structured vacuum energy. This was not a theoretical belief—Tesla built devices that supported this idea. His wireless energy transmission experiments suggested that energy could be extracted directly from space without the need for conventional fuel. He believed that, if properly understood, this hidden energy could power the world indefinitely.

Tesla's claims were not entirely without scientific precedent. Today, modern physics acknowledges that empty space is not truly empty. Quantum field theory suggests that space is teeming with virtual particles and fluctuating energy fields. Dark energy, an unknown force driving the expansion of the universe, also hints at the possibility that space itself contains vast amounts of untapped energy. Tesla had proposed such an idea over a century ago, yet his insights were ignored.

Einstein's relativity left no room for Tesla's energy-based universe. It eliminated the concept of a physical medium for light and gravity, treating space as nothing more than a mathematical structure. Tesla saw this as a step backward. He believed that understanding the hidden structure of space was the key to unlocking the next generation of energy systems. In his eyes, relativity was not just incomplete—it was a distraction from the real discoveries that could revolutionize humanity's use of energy.

How Tesla's Experiments with Wireless Energy Transmission Contradicted Einstein's Space-Time Model

Tesla was not content with theories—he built experiments to test his ideas. His work on wireless energy transmission directly contradicted Einstein's space-time model. If space were truly empty, Tesla argued, then how was he able to transmit energy wirelessly over long distances? His high-voltage experiments demonstrated that energy could move through space in ways that modern physics could not explain.

One of Tesla's most ambitious projects, the Wardenclyffe Tower, was designed to prove these principles. His goal was to transmit electricity wirelessly across the planet, using the structured energy of space itself as a medium. Tesla believed that energy could be sent anywhere in the world without the need for power lines, cables, or fuel. If successful, his technology could have provided limitless free energy to humanity.

But the project was abruptly shut down. Funding was pulled. His patents were buried. His work was dismissed as fantasy. Was Tesla mistaken, or did his discoveries threaten the emerging energy industry? Was he merely ahead of his time, or was he deliberately silenced because his ideas challenged the economic power structures built around fossil fuels?

Tesla's experiments suggested a reality that Einstein's equations did not account for. If space were truly just an abstract mathematical fabric, then Tesla's wireless power transmission should have been impossible. Yet, his experiments showed that energy could be harnessed from space itself. This directly contradicted Einstein's model of an empty, curved space-time.

The Ignored Legacy: Tesla's Insights into Structured Vacuum Energy That Physics Refuses to Acknowledge

Today, Tesla is remembered as a genius inventor, but not as a serious physicist. His contributions to technology are celebrated, yet his deeper insights into space and energy are ignored. His rejection of relativity is often dismissed as a mistake rather than a legitimate scientific challenge.

Yet, modern physics is slowly rediscovering what Tesla already knew. Quantum field theory now suggests that empty space is full of virtual particles, echoing Tesla's structured vacuum energy. Dark energy is now believed to permeate the universe, much like Tesla's concept of an energy-filled space. Zero-point energy research is beginning to explore the possibility of extracting energy from the vacuum—an idea Tesla had proposed over a century ago.

Tesla's rejection of relativity was not ignorance—it was insight. He saw space not as an abstract mathematical concept, but as a real, structured energy system. He understood that electromagnetism and gravity were connected, long before modern physicists began searching for a unified field theory. He recognized that energy was all around us, waiting to be harnessed—at a time when mainstream physics was not ready to accept it.

If Tesla had been taken seriously, physics might have evolved in a completely different direction. We might have developed energy systems based on space energy instead of fossil fuels. We might have unified gravity and electromagnetism long before today's fragmented attempts at a grand unified theory. We might have built a model of the universe based on energy structures rather than abstract space-time geometry.

Tesla was not wrong. He was simply ahead of his time. Einstein's relativity became the dominant model, while Tesla's energy-based vision was ignored. Today, physics still struggles with contradictions that Tesla's ideas might have resolved. The future of science may depend on rediscovering what Tesla saw all along.

The real question is: **Are we ready to listen this time?**

Next: Why Einstein Might Have Agreed But Never Admitted It

- How Einstein's later work hinted at the need for an underlying energy field.
- His struggles with quantum vacuum energy and the limitations of space-time curvature.
- The possibility that Einstein saw Tesla's ideas as valid but could not integrate them into relativity.

Why political and academic pressure may have prevented Einstein from openly challenging his own work.

Because the truth is simple:

Einstein may have been forced to reject Tesla's ideas—not because they were wrong, but because they could not fit within the physics he helped create.

8.3 Why Einstein Might Have Agreed But Never Admitted It

Albert Einstein revolutionized physics by introducing the concept of space-time, redefining gravity as the bending of this four-dimensional fabric. His work replaced Newton's absolute space and time with a dynamic continuum, laying the foundation for modern cosmology, black holes, and quantum mechanics. He became the face of theoretical physics, his name synonymous with genius.

But beneath this towering legacy, a problem remained—one that Einstein himself may have recognized but could never openly admit. As he delved deeper into his own theories, inconsistencies began to surface. He spent decades searching for a unified field theory, suggesting that he knew something was missing from relativity. He struggled with the contradictions between relativity and quantum mechanics, but he never resolved them.

Could it be that Einstein secretly agreed with Tesla's idea of a structured energy field but found no way to integrate it into his own work? Did he suspect that space-time was an incomplete model? Did academic and political pressures prevent him from challenging the very system he had built? This section explores the possibility that Einstein saw the flaws in relativity but was trapped by the weight of his own success.

How Einstein's Later Work Hinted at the Need for an Underlying Energy Field

Einstein was not satisfied with relativity alone. He spent the last decades of his life searching for a deeper framework—a unified field theory that would connect gravity with electromagnetism. He was convinced that the forces governing the universe were linked at a fundamental level, but he could never fully explain how.

If space-time curvature alone was enough to explain gravity, why did Einstein keep searching? If relativity was a complete theory, why did he feel the need to go beyond it? These questions suggest that even Einstein himself suspected that space was not simply an abstract mathematical structure but something with intrinsic properties beyond just curvature.

His later work even touched on ideas uncomfortably close to Tesla's structured energy universe. He explored the concept of an "aether," a medium he had previously dismissed. He investigated equations that suggested space had active, physical properties. He wrestled with the fact that relativity could not explain the fundamental link between gravity and electromagnetism.

In many ways, Einstein was inching toward the very ideas Tesla had advocated all along. He may not have been ready to accept Tesla's conclusions, but his later research suggests that he sensed the limitations of his own model. The very thing Tesla had insisted upon—the idea that space was not empty but filled with structured energy—was something Einstein may have begun to suspect as well.

His Struggles with the Quantum Vacuum and the Limitations of Space-Time Curvature

One of the biggest challenges Einstein faced was the rise of quantum mechanics. His own equations suggested that space could not be truly empty, yet relativity treated space as a passive stage, bending under gravity but devoid of intrinsic energy. Meanwhile, quantum field theory introduced the concept of vacuum energy—the very structured energy Tesla had spoken about.

This created an impossible contradiction. In quantum mechanics, space was filled with fluctuating virtual particles and measurable energy fields. In relativity, space was just an empty fabric that could curve but had no underlying substance. These two pictures of the universe did not fit together.

Einstein knew that space-time curvature alone could not explain everything. His equations broke down in extreme conditions like black holes. They failed to connect with the quantum world. They treated space as a geometric abstraction rather than a dynamic, energy-filled entity.

Was this why Einstein spent his final years searching for something deeper? Did he realize that relativity had ignored a fundamental aspect of space? Did he suspect that Tesla's structured vacuum energy was more than just a wild theory? If so, he never publicly admitted it.

The Possibility That Einstein Saw Tesla's Ideas as Valid But Could Not Integrate Them Into Relativity

Relativity was built on the rejection of the ether—the idea that space was filled with a medium that enabled the transmission of light and energy. But what if Einstein later realized that this rejection had been premature? Tesla had always argued that space was not a void but a structured energy field. If he was right, it meant that relativity was missing a crucial element.

The problem was that relativity was purely geometric. It described motion and curvature but did not account for underlying energy interactions. Tesla's view of space as an active, structured medium would have required a complete overhaul of relativity's foundations.

Einstein could not simply abandon relativity. His entire career, reputation, and scientific legacy were built on it. To admit that space was not empty—that it had intrinsic energy properties—would have been to contradict the very model he had spent his life defending.

Even if Einstein suspected that relativity was incomplete, he had no clear way to fix it. His equations worked mathematically, but they did not connect with the quantum world. If Tesla's structured energy field was real, integrating it into relativity would have required a fundamental rewrite of the theory. And by the time Einstein may have realized this, relativity had already been accepted as the cornerstone of modern physics.

The Political and Academic Pressure: Why Einstein Could Not Openly Challenge the System He Helped Build

By the time Einstein began to question relativity's completeness, it was already too late. The scientific establishment had fully embraced space-time as the foundation of physics. Quantum mechanics had branched off into its own field, further deepening the divide. And Einstein himself had become an icon—his theories treated as unquestionable truth.

Had Einstein publicly admitted that relativity was incomplete, the consequences would have been enormous. His credibility would have been attacked. The physics community would have resisted the idea. The institutions that supported his work would have faced a crisis. The entire framework of modern physics was built on relativity—challenging it would have meant challenging an entire system.

Instead of rejecting relativity, Einstein kept working quietly, searching for something deeper. He never dismissed quantum field theory, but he also never fully accepted it. He never abandoned relativity, but he also never found a way to make it complete. He spent his final years trying to bridge the gap between space-time and energy, but he never succeeded.

What if Einstein had collaborated with Tesla instead of working against his ideas? Tesla had the experimental proof that space contained energy. Einstein had the mathematical genius to refine the theory. Together, they might have built a new model of physics—one that did not rely on space-time patches but on a true understanding of structured energy.

But history played out differently. Relativity was accepted, even with its flaws. Tesla's structured vacuum energy was ignored. And modern physics remains stuck in the contradictions that Einstein never resolved.

Next: 8.4 How Modern Physics Ignored Tesla's Energy-Based Theories

- The shift from energy-based physics to mathematical abstractions.
- How Tesla's vision of structured space was buried by quantum mechanics.
- The consequences of rejecting energy-based physics in favor of space-time models.

Why revisiting Tesla's energy-centric approach could fix modern physics.

Because the truth is simple:

Einstein may have realized Tesla was right—but by then, it was too late to admit it.

8.4 How Modern Physics Ignored Tesla's Energy-Based Theories

The history of physics is filled with competing ideas—some embraced, others discarded. Einstein's space-time model replaced Newton's absolute space and time, and quantum mechanics reshaped our understanding of particles and forces. But Tesla's vision of an energy-filled universe, where space itself was an active medium, was dismissed and forgotten.

Was this a natural evolution of scientific thought, or was it the suppression of ideas that did not fit the new paradigm? Why did physics move away from exploring energy interactions and shift toward mathematical abstractions? How did Tesla's structured vacuum energy theory get buried under the rise of relativity and quantum mechanics? And most importantly, what have been the consequences of rejecting an energy-based approach in favor of abstract space-time models?

This section uncovers how modern physics turned away from Tesla's insights—and why correcting this mistake could be the key to advancing science today.

The Shift from Exploring Energy Interactions to Accepting Mathematical Abstractions

Physics was once a discipline grounded in direct observation, experiment, and intuitive reasoning. Newton's laws were based on measurable motion and force. Maxwell's equations described electromagnetism as energy flows through space. Tesla's experiments demonstrated that space itself contained energy that could be tapped.

But as the 20th century progressed, physics began to move in a different direction. Instead of asking *how* nature works, physicists began focusing on *what equations best describe it.* Instead of searching for the physical nature of forces, they built models that simply predicted behavior. Instead of treating energy as the foundation of reality, they created theories that ignored its deeper nature in favor of elegant mathematical frameworks.

This was a turning point. Einstein's relativity described gravity as space-time curvature but never explained what space actually is. Quantum mechanics treated energy as discrete packets but failed to describe the underlying structure of space. These models were highly successful in predicting results, leading physicists to accept them as reality—even when they lacked a true physical explanation.

Tesla's ideas did not fit into this new approach. He believed that energy, not abstract geometry, was the fundamental building block of the universe. He saw space not as an empty void but as an active medium of structured energy. His experimental discoveries suggested that physics was missing something fundamental.

But instead of being explored, his theories were pushed aside. They contradicted the idea of space as a passive background. They did not align with Einstein's space-time model. And they challenged the growing trend of mathematical physics, which was rapidly becoming the dominant paradigm.

Tesla was not ignored because he was wrong—he was ignored because his ideas did not fit the direction physics had chosen.

How Tesla's Vision of Structured Space Was Buried Under the Rise of Quantum Mechanics

In the early 20th century, two competing views of space existed. Tesla saw space as filled with structured energy—an active medium enabling motion, force, and electromagnetism. Einstein saw space as a mathematical construct—an empty stage where objects move based on gravity. Then came quantum mechanics, which introduced a third perspective:

At the microscopic level, space was no longer empty. Quantum field theory revealed that "empty" space teemed with energy fluctuations and virtual particles. The vacuum was not truly nothing—it had structure, just as Tesla had described.

But rather than recognizing Tesla's insights as an early understanding of quantum vacuum energy, mainstream physics rejected his work. Tesla's structured space implied the existence of a medium—a modern form of the ether that Einstein had supposedly eliminated. His energy-based perspective was rooted in experiment, while quantum mechanics was built on statistical probabilities and uncertainty.

Tesla's work threatened to bridge the gap between classical physics and the emerging quantum world. But instead of investigating his claims, physics ignored them.

If Tesla's structured energy model had been taken seriously, modern physics might have evolved differently. We might have discovered a unified energy framework much earlier. We might have avoided the disconnect between relativity and quantum mechanics. We might not be relying on mathematical fixes like dark matter and dark energy to explain gaps in our models.

Instead, physics buried Tesla's work and moved forward with incomplete theories.

The Consequences of Rejecting Energy-Based Physics in Favor of Space-Time Models

Today, physics faces deep contradictions that both Einstein and Tesla foresaw. Relativity and quantum mechanics remain fundamentally incompatible. Gravity has never been successfully unified with the other forces. Entire new concepts, like dark matter and dark energy, have been invented to explain missing pieces in our models—despite having no direct experimental proof.

These problems suggest something is fundamentally wrong. Physics still lacks a clear definition of space. We do not fully understand energy, despite its role in every interaction. We rely on abstract constructs—quarks, space-time curvature, and force-carrying particles—without knowing their true physical nature.

Tesla's rejection of relativity was not an act of ignorance—it was a warning. He saw that physics was drifting away from reality and into abstraction. He knew that energy, not geometry, was the key to understanding the universe. He tried to demonstrate this with his experiments, but physics refused to listen.

Now, modern physics is paying the price for that decision. Instead of questioning its assumptions, it keeps adding more theoretical constructs to fix problems. It continues treating space as an empty void, despite quantum mechanics proving otherwise. It searches for a "theory of everything" while refusing to reconsider the energy-based framework Tesla proposed over a century ago.

Ignoring Tesla was not just a mistake of the past—it remains a mistake that is still holding science back today.

The Path Forward: Why Revisiting Tesla's Energy-Centric Approach Could Fix Modern Physics

The future of physics does not lie in more mathematical patches. It lies in rediscovering energy as the foundation of reality. We must rethink space—not as an empty void, but as a structured energy system. We must redefine gravity—not as curved space, but as an effect of energy gradients. We must unify the forces—not by adding new particles, but by recognizing energy's fundamental role in all interactions.

Tesla's insights may hold the key. His belief in structured vacuum energy aligns with modern discoveries in quantum fields. His vision of an energy-driven universe could help bridge the gap between relativity and quantum mechanics. His experiments hinted at the possibility that energy could be extracted from space—an idea that may lead to breakthroughs in future energy technologies.

To move forward, physics must acknowledge that relativity was never the full picture. It must recognize that quantum vacuum energy is real and structured, just as Tesla described. It must revisit Tesla's experiments—not as outdated ideas, but as the missing foundation for a new energy-based theory of physics.

Einstein's relativity was an important step—but it was never the final answer. It described gravity but did not explain it. It connected space and time but ignored energy's deeper role. It created a mathematical model but left fundamental questions unresolved.

Tesla's energy-centric approach was not just an alternative—it may have been the correct path all along. The biggest question now is:

Will physics finally correct its course, or will it continue ignoring the insights that could have changed everything?

Because the truth is simple:

Tesla wasn't wrong. Physics just wasn't ready to listen.

Key Takeaways

- **Einstein and Tesla had fundamentally different views on physics**, with Einstein focusing on space-time geometry and Tesla on structured energy fields.
- Einstein's **General Relativity** defined gravity as a warping of space-time, while Tesla argued that **space itself is an active, structured energy medium**.
- Tesla believed the universe operates on **energy interactions, not force-based mechanics**, challenging the need for concepts like gravity as a separate force.
- **Physics suppressed Tesla's work**, favoring Einstein's mathematical models despite lacking a deeper understanding of energy's role in shaping reality.
- **Modern physics still struggles to unify relativity and quantum mechanics**, a problem Tesla's structured energy approach might have resolved.
- The focus on **mathematical abstractions** has prevented physics from fully exploring the possibility that space itself is a **dynamic energy field, not an empty void**.

What If...? (Speculative Exploration)

- What if physics had followed **Tesla's ideas on structured energy** instead of Einstein's space-time model?
- What if **space is not an empty fabric that bends**, but an energy-filled structure that interacts with matter?
- What if Tesla's energy-based physics **could unify gravity and electromagnetism**, something Einstein never achieved?
- What if the suppression of Tesla's work was **not just scientific bias, but also due to political and economic interests**?

Why Did Physics Choose Einstein Over Tesla?

- Einstein's **relativity was mathematically elegant**, making it easier to integrate into existing physics.
- Tesla's approach **lacked mathematical formalism**, making it harder to accept in academic circles.
- The scientific establishment favored **equation-driven physics** over Tesla's **experimental and engineering-based approach**.
- Tesla's ideas on **free energy and structured space** were seen as a **threat to established industries**.
- Einstein's relativity fit well with the **emerging quantum mechanics**, despite leaving gravity disconnected from other forces.

Challenges & Open Questions

- If Tesla was right, **how do we experimentally prove space is an energy medium?**
- Could an **energy-based model of space explain gravity** without needing space-time curvature?
- How do we reconcile **Tesla's structured energy ideas** with modern quantum mechanics?
- Can Tesla's insights help **fix the contradictions between relativity and quantum physics?**

Suggested Readings & References

- **Tesla's Vision of an Energy-Based Universe** – Why Tesla rejected Einstein's relativity.
- **The Divide Between Experimental and Theoretical Physics** – How Tesla and Einstein represented two opposing schools of thought.
- **Chavan's Grand Unified Theory of Physics** – A deeper exploration of structured energy fields as an alternative to space-time curvature.

Part 2: The New Framework – A Universe without Fixes

9: Structured Vacuum Energy

The True Fabric of Reality

For more than a century, physics has been caught in a paradox. It treats space as empty while simultaneously filling it with invisible forces, virtual particles, and mathematical constructs. Relativity describes space as a void that bends under gravity but never explains what space actually *is.* Quantum mechanics treats empty space as teeming with energy fluctuations but refuses to call it a structured medium. Dark matter and dark energy were invented as placeholders to explain missing forces—yet they remain undefined, revealing the cracks in our current models.

In the first part of this book, we exposed the patchwork solutions that turned physics into a web of contradictions. Einstein's relativity abandoned the idea of a structured space but struggled to unify with quantum mechanics. Quantum physics reintroduced vacuum energy but failed to connect it with gravity and space-time. Meanwhile, Tesla's vision of an energy-filled space was ignored—yet modern physics is, unknowingly, circling back to his ideas.

Now, it is time to move beyond these contradictions. What if space is not empty, but structured energy itself? What if gravity, mass, and charge are not independent forces but emerge from energy gradients? What if the universe is not merely a collection of particles but an interconnected energy system?

This chapter introduces the missing foundation of physics—**Structured Vacuum Energy.** It reveals why space is not a void but an active, structured medium. It explains how energy gradients naturally give rise to mass, charge, and forces—eliminating the need for arbitrary "fundamental particles." And it shows how redefining space as structured energy resolves the biggest conflicts in modern physics.

The illusion of empty space ends here. Physics does not need more patches—it needs a new foundation. And that foundation is **Structured Vacuum Energy—the true fabric of reality.**

9.1 The End of the Empty Space Myth – Why Space Is Not Nothing

For centuries, science has struggled to define the nature of space. Newtonian physics treated space as an absolute, empty stage where objects moved. Einstein's relativity replaced this with a flexible fabric that bends under gravity, but it still left space itself undefined. Meanwhile, quantum mechanics shattered the idea of emptiness altogether, revealing that space is filled with fluctuating energy.

These views contradict each other, yet modern physics pretends they can coexist. Relativity describes space as a void, while quantum mechanics describes it as an active energy field. General relativity ignores vacuum energy, while quantum field theory depends on it. Instead of addressing these contradictions, physicists have invented dark matter and dark energy to cover the gaps, rather than questioning their fundamental assumptions.

The truth is simple: space is not empty. It is structured energy. Physics abandoned this idea for decades but was ultimately forced to bring it back. The concept of structured vacuum energy resolves the conflict between relativity and quantum mechanics. And once we accept this, we no longer need artificial fixes like dark matter and dark energy to explain the missing forces in the universe. This section exposes the myth of empty space and reveals why structured vacuum energy is the missing foundation of physics.

How Physics Abandoned the Structured Vacuum but Was Forced to Bring It Back

At the end of the 19th century, scientists believed space had structure. The "ether" was thought to be an invisible medium that carried light waves, an idea that fit naturally with Maxwell's equations of electromagnetism. This suggested that space was not a void but an active, structured energy system.

Then came Einstein. His relativity model eliminated the need for an ether, replacing it with the concept that space itself could bend and stretch. This became the foundation of modern physics—but at the cost of creating new problems. Without a structured vacuum, there was no clear explanation for how light propagated across space. Gravity became a purely geometric effect with no physical mechanism. And quantum mechanics, which required an energy-filled space, became fundamentally incompatible with relativity.

Despite abandoning the structured vacuum, physics was eventually forced to bring it back. Quantum field theory revealed that space was not truly empty but filled with fluctuating energy. The Casimir effect demonstrated that vacuum energy exerted measurable forces. The concept of vacuum energy became undeniable, but instead of calling it a structured space, physicists rebranded it under new terms: zero-point energy, quantum vacuum fluctuations, and dark energy.

This contradiction sits at the heart of modern physics. Space cannot be both "empty" in relativity and "filled with energy" in quantum mechanics. The refusal to acknowledge structured vacuum energy is why physics remains divided. It is time to stop treating space as a paradox and recognize it for what it truly is—an active, structured energy field.

The Contradiction Between Relativity's "Empty Space" and Quantum Mechanics' "Energy-Filled Vacuum"

Modern physics is built on two opposing views of space. Relativity treats space as a geometric void that bends under gravity, while quantum mechanics treats space as an energy field teeming with fluctuations. These two perspectives cannot both be correct.

If space is truly empty, quantum energy fluctuations should not exist. If space is truly filled with energy, relativity's bending "nothingness" makes no sense. Yet physics continues to operate as if both views are simultaneously valid, even though they are fundamentally opposed.

This contradiction has led to a crisis in physics. General relativity cannot explain why quantum vacuum energy exists. Quantum mechanics cannot explain why gravity does not emerge naturally from energy fields. Instead of resolving this, physicists have created artificial constructs—dark matter and dark energy—to hide the problem.

But what if space is not empty, but a structured energy medium? What if gravity is not the bending of nothingness, but an effect of energy density gradients? What if mass, charge, and forces emerge from variations in this structured vacuum? By rejecting the outdated notion of empty space, we can finally bridge the gap between relativity and quantum mechanics.

How Vacuum Energy Is the Missing Link in Unifying Gravity and Quantum Mechanics

One of the biggest failures of modern physics is its inability to unify gravity with quantum mechanics. Relativity describes gravity as curved space-time but does not explain why mass causes curvature. Quantum mechanics describes all other forces in terms of energy fields, yet gravity remains separate. For decades, physicists have searched for a "quantum gravity" theory, but every attempt has failed.

The reason is simple: they are treating space as a void instead of an energy structure. Instead of recognizing gravity as an energy effect, they are searching for hypothetical force carriers like gravitons. Instead of acknowledging the structured vacuum, they continue to cling to the illusion of empty space.

If we accept that space is structured vacuum energy, then gravity is no longer a mysterious geometric effect—it is an interaction of energy fields. Mass is no longer an intrinsic property—it is an emergent effect of energy density.

And most importantly, quantum mechanics and relativity naturally merge, because they are both describing the same underlying energy structure.

Vacuum energy is the missing link. It explains why particles emerge from energy fluctuations. It explains why space exhibits force-like behavior despite being "empty." And it eliminates the artificial divide between relativity and quantum mechanics. This is the foundation that physics has been searching for.

Why Structured Vacuum Energy Replaces the Need for Dark Matter and Dark Energy

One of the most glaring failures of modern physics is the invention of dark matter and dark energy. Dark matter was proposed to explain why galaxies move in ways that defy gravity models. Dark energy was introduced to explain why the universe's expansion is accelerating. Yet neither has ever been directly observed.

But what if dark matter and dark energy are just misinterpretations of structured vacuum energy? What if the missing gravity attributed to dark matter is actually an interaction between vacuum energy and mass? What if dark energy is just a manifestation of large-scale energy density variations? What if both concepts are illusions caused by misunderstanding space as an empty void?

By recognizing structured vacuum energy, we no longer need these artificial constructs. The missing gravity attributed to dark matter can be explained as an interaction between mass and vacuum energy fields. The accelerating expansion of the universe can be understood as an effect of large-scale energy flow. There is no need for "invisible matter" or "mysterious repulsion"—just a deeper understanding of space as structured energy.

This is the final step in replacing the broken models of modern physics. No more patches. No more invisible substances. No more contradictions between relativity and quantum mechanics. The structured vacuum is the key to understanding reality.

Next: Energy Gradients – The True Source of Mass, Charge, and Forces

- Why mass is not an intrinsic property but an energy interaction.
- How charge emerges from structured vacuum fluctuations.
- Why forces are not fundamental—they are effects of energy density variations.
- The bridge between quantum mechanics and gravity through energy gradients.

Because the truth is simple: **Space is not empty—it is structured energy.**

9.2 Energy Gradients – The True Source of Mass, Charge, and Forces

Physics has long treated mass, charge, and forces as fundamental properties of nature—fixed traits assigned to particles without a deeper explanation. Mass is considered an intrinsic characteristic of matter, charge is assigned arbitrarily to particles like electrons and protons, and forces are categorized as fundamental interactions.

But these assumptions stem from a flawed understanding of space and energy. Mass is not an independent property—it is an effect of energy interactions. Charge is not an isolated trait—it emerges from fluctuations in the structured vacuum. Forces are not separate entities—they arise from energy gradients in space.

In reality, energy is the true foundation of everything. Mass is an energy density effect, not an inherent trait of particles. Charge results from energy variations within the vacuum, not a predefined property. Forces emerge naturally from differences in energy distribution, eliminating the need for "force carriers" like bosons.

This section reveals how energy gradients shape reality—and why understanding them is the key to unifying physics.

Why Mass Is Not an Intrinsic Property but an Energy Interaction

Modern physics treats mass as an absolute property of particles. The Higgs mechanism was introduced to "give" mass to particles, but it only describes how mass behaves, not why it exists. General relativity links mass to space-time curvature, yet it never explains why mass affects space in the first place. Quantum mechanics assigns mass to particles but does not define what mass truly is.

This approach raises more questions than answers. Why does mass exist at all? Why does it interact with gravity? Why do different particles have different masses?

The answer lies in structured vacuum energy. Mass is not an intrinsic feature—it is an effect of energy density. Particles are stable energy configurations, and their "mass" is a measure of their interaction with the surrounding energy field. A high-energy-density region of space behaves like mass, influencing surrounding energy gradients and creating gravitational effects.

This means mass is not a built-in property—it is a dynamic result of energy interactions. If structured vacuum energy changes, mass also changes—mass is an effect, not a cause. Gravity does not pull objects—it is simply the result of energy density differences guiding motion.

Understanding mass as an emergent property of vacuum energy removes the need for artificial constructs like the Higgs field. Mass is just energy in a particular configuration. There is no "mass-giving" mechanism—only energy structure variations. If vacuum energy density shifts, mass can change, explaining effects like mass-energy equivalence and fluctuations in particle behavior.

This is a completely new way to understand mass—not as a property of matter, but as an interaction of energy.

How Charge Emerges from Structured Vacuum Fluctuations

Charge is another mystery in physics. Electrons are assigned a charge of -1, protons +1—but why? Charge is treated as fundamental, yet it has no deeper explanation. The force between charged particles is described mathematically, but never explained physically.

The standard model treats charge as a given—a property assigned to particles like a label. But if space is not empty, charge must emerge from interactions with structured vacuum energy.

Vacuum energy fluctuations create imbalances, leading to charge differences. Positive and negative charges are opposite energy states, not intrinsic properties. Charge is not a fundamental trait of matter—it is a property of how particles interact with the structured vacuum.

This explains why opposite charges attract—they represent complementary energy imbalances that naturally restore equilibrium. It also explains why charge is quantized—because energy density structures only allow discrete charge states. Charge conservation emerges naturally because vacuum energy fluctuations maintain overall energy balance.

Charge is not a built-in feature of particles—it is an effect of their interaction with vacuum energy. When we redefine charge as an energy field effect, it no longer needs arbitrary assignments—it emerges naturally from structured vacuum interactions.

Why Forces Are Not Fundamental – They Are Effects of Energy Density Variations

Physics treats forces like gravity, electromagnetism, and nuclear interactions as fundamental. The Standard Model describes them as separate forces, each carried by different particles. Gravity is treated as space-time curvature, separate from the other forces. Electromagnetism is mediated by photons, while nuclear forces rely on gluons and bosons.

But what if forces are not independent at all? What if they are all just effects of energy density variations in the structured vacuum?

Consider this: Gravity arises from large-scale energy gradients. Electromagnetism emerges from charge-induced energy variations. The nuclear forces are just short-range effects of intense energy density differences.

This means there are no separate "fundamental" forces—only different manifestations of energy density interactions. Particles do not "exchange force-carrying bosons"—they interact through vacuum energy gradients. All forces are unified under a single concept—variations in structured vacuum energy.

This explains why gravity is so weak compared to electromagnetism—it is an effect of large-scale energy balance, not a direct force. It also explains why nuclear forces have limited range—they are caused by localized high-energy-density regions. Force carriers like photons, gluons, and bosons are unnecessary—they are just mathematical placeholders for energy interactions.

Forces are not fundamental. They are just different effects of how energy is distributed.

The Bridge Between Quantum Mechanics and Gravity Through Energy Gradients

The greatest challenge in physics has been unifying quantum mechanics and gravity. Quantum mechanics works at small scales but ignores gravity.

Relativity describes gravity but does not fit with quantum mechanics. No successful "quantum gravity" theory has been found.

The reason? Physics is still treating forces as separate things instead of emergent effects of energy.

When we recognize that energy gradients govern all interactions, the problem disappears. Gravity is just a large-scale energy effect, not a separate force. Quantum mechanics describes small-scale energy interactions, which naturally extend to gravity. All forces become part of a unified system—interactions within structured vacuum energy.

This means there is no need to "quantize" gravity—gravity is already an emergent effect of quantum energy density variations. There is no need for exotic force carriers—forces are just how energy gradients influence motion. There is no separation between gravity and quantum mechanics—both are describing different scales of the same energy framework.

By recognizing that mass emerges from energy density, charge arises from structured vacuum fluctuations, and forces are energy density variations, we naturally unify physics under a single framework—structured vacuum energy.

Next: The Universe as an Energy System, Not a Particle System

- Why the Standard Model's particle approach is an illusion.
- How structured energy fields naturally give rise to the behavior of matter.
- Why particles are stable energy configurations, not discrete objects.
- How rethinking the universe as an energy system resolves the contradictions in physics.

Because the truth is simple: **Mass, charge, and forces are not fundamental—they are effects of energy gradients.**

9.3 The Universe as an Energy System, Not a Particle System

For over a century, physics has treated the universe as a collection of discrete particles interacting through fundamental forces. The Standard Model categorizes reality into quarks, leptons, and bosons, each assigned specific properties. Relativity treats mass as an intrinsic characteristic, warping space-time. Quantum mechanics introduces uncertainty, describing particles as probability clouds rather than concrete objects.

But this particle-based approach raises more questions than it answers. It requires force-carrying particles (bosons) to explain interactions, yet these have never been directly observed in isolation. It treats mass and charge as fundamental properties without revealing their true origin. It divides physics into incompatible models—quantum mechanics and relativity—because particles and space-time do not fit together in a unified theory.

What if particles are not the fundamental building blocks of the universe? What if they are simply stable energy formations within a structured vacuum? What if space is not a passive void but an active energy field that gives rise to all physical properties? What if forces emerge naturally from variations in energy density, rather than from the exchange of "force-carrying" particles?

This section exposes the illusion of a particle-based universe and explains why energy—not particles—is the true foundation of reality.

Why the Standard Model's Particle Approach Is an Illusion

For decades, the Standard Model has been treated as the most accurate description of nature. It divides matter into fermions (quarks and leptons) and force carriers (bosons), explaining interactions through quantum field theory, where particles exchange force-mediating particles. Whenever equations fail to match observations, new "missing particles" are proposed to fill the gaps.

But this approach has fundamental problems. Quarks are said to be the building blocks of protons and neutrons, yet they are never observed in isolation. Gluons supposedly bind quarks together, but they cannot be detected as individual particles. The Higgs boson was introduced to "give" mass to particles, yet it does not reveal the true nature of mass.

Physics has become dependent on theoretical constructs rather than physical explanations. The Higgs boson was not discovered as a standalone entity—it was inferred based on statistical deviations in experimental data. The W and Z bosons were introduced to explain weak interactions, yet their existence remains tied to mathematical necessity rather than empirical certainty.

This endless cycle of particle-hunting has led physics into a dead end. Instead of simplifying our understanding, it has made physics unnecessarily complex. Instead of unifying nature, it has divided physics into a patchwork of unexplained interactions. Instead of looking for deeper principles, physicists keep adding new entities to force equations to work.

But what if particles are not fundamental? What if they are merely stable configurations of energy within a structured vacuum? Shifting our perspective from particles to energy fields resolves these contradictions.

How Structured Energy Fields Naturally Give Rise to the Behavior of Matter

If space is a structured vacuum filled with energy, then:

- What we call "particles" are simply energy concentrations in stable formations.
- Forces are not particle exchanges—they are natural effects of energy density variations.
- Mass, charge, and spin emerge from energy interactions rather than being built-in properties of particles.

Consider an electron. In the Standard Model, it is treated as a fundamental particle with a negative charge. Its charge and mass are assigned arbitrarily, without a deeper explanation. Quantum mechanics describes its behavior probabilistically but never defines what an electron truly is.

Now, imagine the electron as a stable energy structure within the vacuum. Its charge is an effect of its interaction with surrounding vacuum energy fluctuations. Its mass is a result of how it distorts local energy density. Its spin is an inherent feature of energy flow within the structured vacuum.

This perspective removes the need for arbitrary particle properties and force carriers.

- Matter is a product of energy interactions, not a collection of independent particles.
- Charge, mass, and forces are natural effects of energy organization, not separate entities.
- Physics moves from fragmented descriptions to a unified understanding of matter and energy.

This model also resolves the mystery of wave-particle duality. Particles are not tiny, solid objects moving through space. They are stable, localized energy configurations interacting with a dynamic energy field. They appear as discrete particles when measured, but their true nature is wave-like within structured space.

Once we stop thinking of particles as physical objects and start recognizing them as energy phenomena, many of physics' deepest contradictions disappear.

Why Particles Are Stable Energy Configurations, Not Discrete Objects

The idea of particles as fundamental objects is a historical misconception. Ancient physics treated atoms as indivisible units—until protons, neutrons, and electrons were discovered. Later, protons and neutrons were split into quarks—yet quarks are never seen in isolation. Now, physicists search for even smaller "fundamental" entities, not realizing they are chasing an illusion.

What if particles are not objects at all? What if they are just stable energy knots in the structured vacuum? What if their properties—mass, charge, and spin—arise from their interaction with surrounding energy fields?

In this model:

- Protons and neutrons are not made of "quarks"—they are stable energy formations.
- Electrons are not "point particles"—they are energy vortexes with specific charge behaviors.
- Neutrinos are not ghost-like entities—they are low-energy disturbances in the structured vacuum.

This explains:

- Why quantum mechanics cannot pinpoint an electron's location—it is an energy wave, not a fixed object.
- Why quarks are never observed alone—because they do not exist as individual entities, only as energy relationships.
- Why forces act at specific ranges—because they are not independent things, but energy gradient effects.

By redefining particles as stable energy formations, we solve the biggest contradictions in physics.

How Rethinking the Universe as an Energy System Resolves the Contradictions in Physics

The current model of physics is broken because it treats particles, forces, and space as separate concepts. Relativity describes space-time but ignores quantum mechanics. Quantum mechanics describes energy fluctuations but ignores gravity. The Standard Model categorizes particles without explaining their true nature.

By shifting to an energy-based framework:

- Particles are no longer treated as objects—they are energy knots within a structured vacuum.
- Forces are no longer independent interactions—they are energy density gradients.
- Space is no longer a void—it is a dynamic energy field that shapes reality.

This resolves:

- The conflict between relativity and quantum mechanics—because both emerge from the same energy framework.
- The confusion over wave-particle duality—because particles are just wave-like energy structures.
- The need for force carriers—because forces arise naturally from energy variations, not from bosons.

Physics must move beyond the particle illusion and embrace the true foundation of reality—energy.

Next: The End of Fundamental Particles – Why Reality Is Purely Energy

- Why searching for smaller "fundamental" particles is an endless loop.
- How energy configurations, not particles, create the properties of matter.
- Why mass, charge, and spin arise from structured vacuum dynamics—not from elementary particles.
- How shifting from a particle-based model to an energy-based model will revolutionize physics.

Because the truth is simple:

Particles are not fundamental—they are stable energy formations within a structured vacuum.

Key Takeaways

- Modern physics contradicts itself by treating space as both **empty (in relativity)** and **full of energy (in quantum mechanics)**, yet it refuses to define what space actually is.
- The **Standard Model and relativity** fail to unify because they ignore the fundamental nature of space—structured vacuum energy.
- Dark matter and dark energy were **invented as placeholders** to explain missing forces, but they remain undefined, exposing the flaws in our current models.
- If space is a **structured energy field**, then mass, charge, and forces **are not fundamental properties** but emergent effects of energy density variations.
- This **revolutionary shift** eliminates the need for force carriers, extra dimensions, and hypothetical particles, unifying physics under a single framework of structured energy.

What If...? (Speculative Exploration)

- What if **space is not a void**, but an active energy system that shapes motion, gravity, and electromagnetism?
- What if mass, charge, and forces **are not independent properties** but just manifestations of energy interactions?
- What if **structured vacuum energy naturally explains dark matter and dark energy**, eliminating the need for hypothetical, undetectable substances?
- What if **quantum mechanics and relativity are not separate laws**, but different views of the same structured energy framework?

Why Has Physics Ignored Structured Vacuum Energy?

- Einstein's **relativity eliminated the ether**, rejecting the idea of structured space, but this **created more contradictions than solutions.**
- Quantum mechanics **accidentally reintroduced vacuum energy**, yet physics refused to acknowledge it as a real structure.
- Instead of questioning the assumption that space is empty, physicists **patched the contradictions with dark matter, dark energy, and quantum fields.**
- The focus on **equation-driven physics** led scientists to prioritize mathematical consistency over physical reality.

Challenges & Open Questions

- If space is structured energy, **what experiment can directly measure its structure?**
- How does structured vacuum energy explain **why mass influences gravity?**
- Could an energy-based approach **eliminate the need for force-carrying particles like gluons and photons?**
- If forces emerge from energy gradients, **what governs their interaction at different scales?**

Suggested Readings & References

- **Tesla's View of an Energy-Based Universe** – Why Tesla rejected the concept of empty space.
- **The Contradictions Between Relativity and Quantum Mechanics** – How structured energy resolves this conflict.
- **Chavan's Grand Unified Theory of Physics** – A deeper explanation of structured vacuum energy as the foundation of physics.

10: The Atomic Truth

Energy Configurations, Not Particles

For over a century, physics has described atoms as miniature solar systems, with electrons orbiting the nucleus like planets around the Sun. Early models treated electrons as tiny, charged particles moving in fixed paths. Quantum mechanics later introduced the idea that electrons behave like waves, existing as "probability clouds" rather than defined objects. The Standard Model complicated the picture further, breaking protons and neutrons into quarks, held together by invisible force-carrying gluons.

But these descriptions are mathematical conveniences—not physical truths. Electrons are not small particles randomly jumping between orbitals. Protons and neutrons are not "bags of quarks" glued together by mysterious forces. Quantum mechanics is not a fundamental law governing nature—it is an emergent effect of energy interactions within structured space.

In the previous chapter, we dismantled the particle illusion, revealing the universe as an energy system rather than a collection of discrete objects. Now, we apply the same principle at the atomic level. Electrons, protons, and neutrons are not particles at all—they are structured energy formations. Quantum behavior is not random—it emerges naturally from interactions within the structured vacuum.

This chapter rewrites atomic physics from the ground up. Electrons are not particles or waves—they are energy zones shaped by the vacuum field. Protons and neutrons are not quark-based objects—they are stable energy knots formed by structured space. Quantum mechanics does not describe an underlying reality—it is simply the statistical effect of deeper energy interactions.

The atomic truth has been buried beneath layers of mathematical fixes. Now, we expose what atoms truly are—configurations of structured energy.

10.1 Electrons: Structured Energy Zones, Not Particles or Waves

For over a century, physics has struggled to define what an electron actually is. Is it a tiny charged particle orbiting the nucleus? Is it a wave-like probability cloud spread across space? Or is it something entirely different—something that has been misunderstood all along?

The traditional view of electrons relies on contradictory descriptions. Classical physics treats the electron as a point particle with a negative charge. Quantum mechanics describes it as a wave function, a cloud of probabilities. In modern atomic theory, it "jumps" between energy levels, yet no one has ever observed it physically traveling between states.

These contradictions exist because physics has been asking the wrong questions. Electrons are not particles moving in orbits. They are not waves floating in uncertainty. They are structured energy zones—stable formations within the structured vacuum. By redefining electrons as energy configurations rather than particles, we eliminate the paradoxes of quantum mechanics and reveal the true nature of atomic structure.

Why the Electron's "Dual Nature" Is a Misunderstanding

One of the most famous concepts in physics is wave-particle duality—the idea that electrons behave as both particles and waves. In some experiments, electrons appear as tiny particles, producing discrete impacts on a detector. In others, they behave like waves, creating interference patterns. This led physicists to conclude that electrons must be both—a strange hybrid of two contradictory behaviors.

But this interpretation is based on a flawed assumption. There is no "duality"—there is only structured energy interacting with its environment. Electrons do not switch between being a wave and a particle; they are always energy structures responding to the conditions around them.

Consider a whirlpool in water. From a distance, it appears as a stable, well-defined structure. But up close, it is nothing more than flowing energy—without fixed edges or defined boundaries. Its behavior depends entirely on how it is observed.

Electrons behave the same way. They are not tiny objects but stable energy formations in the structured vacuum. Their interactions with surrounding energy fields create the effects we interpret as wave-particle duality. The wave function does not describe an electron as a probability—it describes its structured energy state. Once we remove the artificial distinction between waves and particles, the electron is revealed for what it truly is—a localized energy configuration within the vacuum field.

How Electrons Exist as Localized Energy Configurations Within the Structured Vacuum

If space is not empty but an active energy field, then electrons are not independent objects floating in isolation. They are stable formations embedded in this structured field, constantly exchanging energy with their surroundings.

This explains several key properties of electrons:

- Why they do not spiral into the nucleus despite their attraction to protons.
- Why they exist in discrete energy levels rather than moving continuously.
- Why they exhibit quantum behavior rather than classical motion.

Electrons are not tiny charged balls circling a nucleus. They are stable patterns of energy shaped by vacuum dynamics. Much like a vortex in a fluid, they maintain their form within a larger energy system. They do not "orbit" the nucleus like planets—they exist as standing energy formations. They do not "jump" between energy levels—they undergo structured energy transitions.

By redefining electrons as structured energy zones, we naturally explain their behavior without resorting to the paradoxes of wave-particle duality or probabilistic interpretations.

Why the Concept of "Orbital Shells" Is Just an Energy Density Effect

Traditional atomic theory portrays electrons as orbiting the nucleus in fixed shells, similar to planets around a star. This image, while useful for early models, does not accurately represent reality.

- Electrons do not travel in fixed paths.
- They do not "jump" between levels like satellites changing altitude.
- They do not exist as physical objects moving through space.

Instead, electrons occupy structured energy zones within the vacuum field. These "shells" are not fixed locations but energy density patterns. Higher energy levels correspond to regions where vacuum energy interactions create stable formations.

This explains:

- Why electron positions are described as probability clouds—because they are not localized objects but dynamic energy structures.
- Why each energy level has a discrete value—because it corresponds to a stable vacuum energy configuration.
- Why electrons appear to "disappear" and "reappear" in quantum jumps—because their energy reorganizes within the vacuum, rather than physically moving between locations.

By replacing the outdated concept of orbital shells with structured energy zones, we see atomic behavior as a natural consequence of vacuum energy interactions. There is no mystery. There is no need for duality. There is only energy, organizing itself into stable patterns.

How Energy Field Interactions Explain Electron Movement Without Particle-Wave Duality

The idea that electrons "move" like classical objects is another misconception.

- Particles move through space—but electrons are not particles.
- Waves spread across space—but electrons are not waves.
- Instead, electrons shift within the structured vacuum, responding to energy gradients.

Think of a ripple on the surface of water. The disturbance moves across the surface, but the water itself does not physically jump from one place to another. This is how electrons behave—they shift within the structured vacuum, responding to energy distributions rather than physically moving through space.

This explains:

- Why electrons do not "orbit"—because their energy zones are already defined by the vacuum field.
- Why they transition instantly between states—because they are reorganizing energy, not traveling through space.
- Why they exhibit quantum tunneling—because their structured energy interacts with the vacuum field, allowing transitions without classical motion.

Once we abandon the idea of electrons as particles, all contradictions disappear.

- No need for hidden variables.
- No need for paradoxes like wave-particle duality.
- No need for virtual particles to explain interactions.

Electrons are structured energy zones—shaped by the vacuum field, defined by energy gradients, and governed by natural stability conditions. This is the atomic truth that modern physics has overlooked.

Next: Protons and Neutrons – Stable Energy Knots, Not Quark-Based Objects

- Why quarks have never been observed as free particles.
- How protons and neutrons form as stable energy formations.
- Why nuclear stability depends on vacuum energy interactions, not gluons.
- How mass and charge emerge naturally from structured energy fields.

Because the truth is simple:

Electrons are not particles or waves—they are structured energy configurations within the vacuum field.

10.2 Protons and Neutrons: Stable Energy Knots, Not Quark-Based Objects

For decades, physics has described protons and neutrons as "bags of quarks" held together by gluons. The Standard Model claims that protons and neutrons are made of three quarks—yet quarks have never been observed in isolation. The strong nuclear force is said to "glue" them together, but the exact mechanism remains undefined. Gluons, the supposed carriers of this force, have never been directly observed as individual particles.

But what if protons and neutrons are not composite objects at all? What if they are stable energy formations, not clusters of invisible particles? What if mass and charge emerge naturally from structured vacuum energy? What if nuclear stability is an effect of energy balance rather than the exchange of hypothetical force-carrying particles?

This section redefines protons and neutrons—not as quark-based structures, but as stable knots of energy within the vacuum field.

Why Quarks Have Never Been Observed as Free Particles

The Standard Model claims that protons and neutrons are made of quarks—three in each, bound together by the strong nuclear force. A proton is supposedly composed of two "up" quarks and one "down" quark, while a neutron is made of two "down" quarks and one "up" quark. These quarks are said to be held together by gluons, the force-carrying particles responsible for the strong interaction.

But there is a major problem with this model. Despite decades of experiments, no one has ever observed a quark in isolation. Every attempt to isolate a quark has failed, and it is always found bound inside a larger structure. To explain this, physicists introduced the concept of "quark confinement"—a mathematical workaround that justifies why quarks can never be seen individually.

If quarks cannot exist alone, then do they truly exist as fundamental particles? Or is the entire quark-gluon model just another artificial patch to fix an incomplete theory? Instead of being independent building blocks, quarks may simply be mathematical descriptions of energy distributions within protons and neutrons. If protons and neutrons are stable configurations of structured vacuum energy, this naturally explains why quarks never appear in isolation—because they are not real objects, but rather patterns of energy interactions.

This also explains why the strong force appears so powerful. Instead of being a separate force mediated by gluons, the strong interaction may simply be a manifestation of vacuum energy effects maintaining the stability of protons and neutrons. Their behavior as unified entities suggests that they are not built from smaller parts at all, but rather exist as indivisible energy formations.

If we abandon the quark-gluon model, we open the door to a much deeper understanding of nuclear structure. We no longer need hypothetical particles to explain stability, nor do we need a "glue" force holding quarks together. Instead, nuclear matter can be understood as stable energy formations governed by structured vacuum interactions.

How Protons and Neutrons Form as Stable Energy Formations

If space is filled with structured vacuum energy, then matter itself is not made of separate particles—it is composed of stable energy configurations. Protons and neutrons are not "bags of quarks"; rather, they are localized energy knots within the vacuum field. Their charge and mass do not come from smaller particles but emerge from the way energy is structured within them.

A useful analogy is a tornado. A tornado appears to be a solid entity, but it is just a swirling pattern of air. It remains stable because of the balance between surrounding pressure and flow, and its energy is concentrated in a specific region, giving it structure and behavior.

Protons and neutrons are the same. They are not clusters of smaller particles, but swirling knots of energy within the structured vacuum. Their charge and mass arise from the way energy moves within them, rather than from intrinsic properties assigned to particles.

This explains why protons and neutrons are incredibly stable. Their structure is reinforced by vacuum energy interactions, making them persistent over time. It also explains why protons and neutrons have nearly identical masses—because they are formed by similar energy configurations, differing only in charge and stability conditions. Neutrons, being slightly less stable, decay outside the nucleus because their energy balance depends on surrounding vacuum interactions.

In this model, protons and neutrons no longer need quarks or gluons to explain their behavior. Their properties emerge from vacuum energy interactions, not from fundamental components inside them. Nuclear physics then becomes a study of energy stability rather than force-mediated interactions between imagined particles.

Why Nuclear Stability Depends on Vacuum Energy Interactions, Not Gluons

The strong nuclear force has always been a mystery. It is described as the most powerful force in nature, yet it only acts at extremely short distances. It is said to hold protons and neutrons together inside the nucleus, yet the exact mechanism of how this works remains unclear. The theory relies on gluons, particles that have never been directly detected.

If protons and neutrons are not quark-based, then the strong force itself may not be necessary. Instead of being "glued" together, protons and neutrons remain together because their energy structures naturally balance within the vacuum field. Nuclear stability is not the result of a force—it is the result of energy interactions maintaining equilibrium.

A useful analogy is a soap bubble. A bubble maintains its shape not because of an internal force, but because of pressure balance in the surrounding air. Its stability comes from interactions with the environment rather than from a glue-like force inside it. The nucleus is similar. It is not held together by gluons but is stabilized by structured vacuum energy.

This perspective explains why nuclear forces act only at short range. The interactions between protons and neutrons are not caused by a separate force but are a natural consequence of energy density effects in the vacuum. When a nucleus is disturbed—such as during radioactive decay—it is not that a force weakens, but rather that the energy balance shifts, allowing the structure to collapse and release energy.

By replacing the strong force with structured vacuum energy interactions, we remove another artificial construct from physics. We no longer need force-carrying gluons or the concept of "color charge" in quarks. Instead of complex mathematical patches, we have a natural explanation rooted in the fundamental properties of energy itself.

How Mass and Charge Emerge Naturally from Structured Energy Fields

The Standard Model treats mass and charge as fundamental properties, but it does not explain why they exist. The Higgs boson is said to "give" mass to particles, yet it does not reveal why mass itself is a property of matter. Charge is treated as an intrinsic trait of protons and electrons, yet no one knows why it has specific values. These properties are accepted as "given" rather than being derived from deeper principles.

If space is filled with structured vacuum energy, then mass is not a fundamental property—it is an effect of energy concentration. Charge is not an inherent trait—it is an effect of energy field interactions. Protons and neutrons are not discrete particles with assigned properties; they are stable configurations of moving energy.

This explains why mass changes with energy interactions, as demonstrated by $E=mc^2$. It also explains why charge is quantized—because vacuum energy fields only allow stable energy states. The different masses of particles arise not from fundamental distinctions but from variations in how each structure interacts with vacuum energy.

Everything in physics becomes a question of energy balance, rather than a list of arbitrary particle properties. Instead of searching for missing particles or force carriers, we can redefine mass and charge as emergent effects of structured vacuum interactions.

Next: The Quantum Illusion – How Quantum Mechanics Emerges from Vacuum Energy Fields

The contradictions of quantum mechanics vanish when we recognize that quantum probability is just an effect of structured vacuum fluctuations. Uncertainty arises not from randomness but from interactions between energy configurations. Wavefunctions describe energy field distributions, not "real" particle behavior. By acknowledging quantum mechanics as an emergent property rather than a fundamental law, we move closer to a unified model of physics.

Because the truth is simple: protons and neutrons are not made of quarks—they are stable energy formations within structured vacuum energy.

10.3 The Quantum Illusion: How Quantum Mechanics Emerges from Vacuum Energy Fields

For over a century, quantum mechanics has been celebrated as the most successful yet strangest theory in physics. Particles behave like waves, yet also appear as discrete objects. Outcomes are probabilistic rather than deterministic, creating an apparent randomness at the foundation of reality. The "observer effect" suggests that measurement influences the state of a system, as if reality itself changes when we look at it.

Physicists have attempted to make sense of these bizarre behaviors with competing interpretations. The Copenhagen Interpretation tells us to "shut up and calculate" rather than explain what quantum mechanics actually means. The Many-Worlds Interpretation proposes that every possible event spawns a new universe, introducing infinite complexity without addressing the underlying mechanisms. Pilot Wave Theory suggests the existence of hidden variables but cannot explain how they work in practice.

But what if quantum mechanics is not mysterious at all? What if its strange behavior is simply the result of interactions within structured vacuum energy? What if uncertainty, probability, and wavefunctions all emerge naturally from energy fluctuations in space? What if quantum mechanics is not a fundamental law, but an emergent property of deeper energy dynamics?

This section reveals the quantum illusion—exposing how quantum mechanics arises from structured vacuum energy fields and why its paradoxes are only a misunderstanding of energy interactions.

Why Quantum Probability Is Just an Effect of Structured Vacuum Fluctuations

Quantum mechanics describes reality through probability. An electron does not have a definite position but instead exists as a probability distribution spread across space. A photon does not take one path but all possible paths, interfering with itself in the process. A particle's state is unknown until measured, collapsing instantly into a defined outcome.

Physics assumes that this probabilistic nature is a fundamental feature of reality. Reality at the smallest scales is thought to be inherently indeterminate, with quantum randomness treated as an unavoidable aspect of nature. Yet, despite being deeply ingrained in physics, no one has ever explained why the universe behaves this way.

The truth is simpler than physicists realize. Quantum probability is not fundamental—it is an effect of structured vacuum fluctuations. Space is not empty; it is an active energy field that constantly interacts with matter. Particles are not independent objects; they are stable energy formations shaped by these interactions.

This naturally explains why quantum probabilities exist. Energy fields fluctuate, influencing the states of particles, creating what we interpret as probabilistic behavior. Quantum superposition occurs not because a particle "exists in multiple states" but because energy distributions shift within the vacuum field. Measurement collapses a quantum system not due to an observer's influence but because interaction with the vacuum field forces one possible energy state to stabilize.

Probability is not an intrinsic feature of reality—it is an illusion created by energy field dynamics. There is no need for multiple universes, no requirement for hidden variables, and no fundamental randomness. The apparent uncertainty in quantum mechanics is simply the result of interactions within an ever-shifting energy landscape.

How Uncertainty Arises from Interactions Between Energy Configurations

Quantum mechanics introduces the idea of uncertainty—the principle that certain properties, such as position and momentum, cannot be measured with complete precision. The Heisenberg Uncertainty Principle states that the more accurately we measure one property, the less precisely we can know the other. This is typically interpreted as an unavoidable limit of nature, a built-in fuzziness at the quantum level.

But why should the universe be fundamentally uncertain? If reality is governed by deterministic laws at larger scales, why would small-scale interactions behave differently? The answer lies in structured vacuum energy.

Particles are not fixed points moving through empty space. They are stable energy formations interacting with the vacuum field, which itself is never static. Measuring a particle disturbs this energy balance, causing shifts in local energy density. These shifts alter the particle's momentum or position, creating the effect we call uncertainty.

A boat floating on a perfectly still ocean can have its exact position and speed measured with high precision. But if the ocean is full of waves, every measurement of the boat's position will be slightly different depending on the motion of the water. The more one tries to pinpoint location, the harder it becomes to measure speed accurately, and vice versa.

This is exactly how vacuum energy fluctuations affect quantum measurements. Particles exist within an ever-changing energy field, meaning their properties are never fixed because the vacuum itself is always in motion. Uncertainty is not a fundamental principle of nature—it is simply an inevitable consequence of structured energy interactions.

Why Wavefunctions Describe Energy Field Distributions, Not "Real" Particle Behavior

Quantum mechanics relies on the wavefunction, a mathematical description of a system's probability distribution. It predicts the likelihood of finding a particle in a certain state, yet physicists debate whether the wavefunction represents reality or is just a calculation tool. Some argue it is an abstract probability wave, while others believe it describes an actual physical entity.

But the real question is: what is the wavefunction actually describing?

If reality is made of structured vacuum energy, then the wavefunction is not a probability wave—it is a representation of energy distributions within the vacuum field. Quantum behavior is not the result of randomness; it is a consequence of how energy interacts with itself.

This explains why electrons appear as probability clouds. They are not particles moving along defined paths, but structured energy formations that exist as distributions in the vacuum. It explains why particles interfere with themselves, because they are interacting with vacuum energy waves rather than behaving as independent objects. It also explains why wavefunction collapse occurs upon measurement—not because reality "chooses" an outcome, but because interaction with the vacuum field forces an energy state into a stable configuration.

Quantum mechanics has long been asking the wrong questions. Particles do not have "hidden positions" waiting to be revealed. Reality is not fundamentally probabilistic—it is a dynamic energy system. Wavefunctions are not mystical abstractions—they are direct descriptions of structured energy interactions.

Once we recognize this, quantum mechanics finally makes sense. The paradoxes of quantum probability vanish. There is no need for hidden variables, no need for alternative universes, and no need for philosophical debates about the "nature of reality." Everything is just energy, interacting with energy.

How Recognizing Quantum Mechanics as an Emergent Property Leads to a Unified Physics Model

The biggest failure of modern physics is the artificial separation between quantum mechanics and relativity. Quantum mechanics describes microscopic particles but cannot explain gravity. Relativity describes gravity but has no connection to quantum mechanics. For decades, physicists have struggled to merge the two, yet every attempt at unification has failed.

But the failure exists because physics is still treating quantum mechanics as a separate law of nature rather than an emergent property.

Quantum mechanics is not an independent framework—it is a consequence of vacuum energy interactions. Once we redefine space as structured energy, the conflict between quantum mechanics and relativity disappears.

This leads to a new, unified physics model. Vacuum energy fields replace the concept of empty space. Particles are not discrete objects but stable energy formations within the vacuum. Forces emerge from energy density variations rather than from the exchange of virtual bosons. Quantum probability and uncertainty arise from energy fluctuations, not from fundamental randomness.

With this framework, quantum mechanics and gravity naturally merge. Gravity is no longer treated as a separate force but as an effect of vacuum energy density gradients. Quantum behavior is no longer seen as strange or paradoxical, because it is simply the result of small-scale energy fluctuations. Mass, charge, and spin are no longer arbitrary properties assigned to particles but emergent effects of structured energy interactions.

Physics does not need more mathematical patches to force compatibility between theories. It needs to recognize that quantum mechanics is not a mystery—it is merely the visible effect of deeper energy dynamics.

This is the path to a true unified theory of nature. Quantum mechanics is not a separate, unexplained phenomenon. It is an illusion created by incomplete models—a veil that disappears when we finally see reality as structured energy.

Key Takeaways

- Atoms are not **collections of discrete particles** but **structured energy configurations** that define their behavior.
- The **electron, proton, and neutron** are not fundamental building blocks but **manifestations of energy density and structured interactions.**
- The **wave-particle duality paradox** vanishes when we stop treating electrons and photons as objects and instead see them as **energy distributions in space.**
- Quantum mechanics **misinterprets atomic behavior** by treating it probabilistically instead of recognizing structured energy formations.
- The **so-called "empty space" inside atoms** is not empty—it is an active energy field that governs atomic stability.
- If atoms are **energy structures, not collections of particles,** this changes our understanding of chemistry, nuclear physics, and even material science.

What If...? (Speculative Exploration)

- What if electrons **do not orbit the nucleus**, but are just **stable energy configurations shaped by surrounding fields**?
- What if protons and neutrons **are not separate particles**, but **localized energy densities interacting within a structured space**?
- What if atoms do not need **quantum probabilities**, but are fully deterministic energy systems governed by precise interactions?
- What if physics abandoned the **particle-based approach**, leading to a new understanding of atomic interactions?

Why Has Physics Treated Atoms as Particle Collections?

- **Early experiments like Rutherford's gold foil experiment** led physicists to assume atoms had a solid nucleus with orbiting electrons.
- The **Bohr model of the atom** simplified atomic behavior mathematically but **did not truly describe what an atom is.**
- Quantum mechanics abandoned **a real physical explanation** and settled for probability-based descriptions.
- The **Standard Model reinforced the particle myth**, adding quarks and gluons instead of questioning the assumptions behind atomic structure.
- **Structured energy models** were never explored in mainstream physics because they challenged the existing framework.

Challenges & Open Questions

- If electrons are not particles, **what exactly are they?**
- How does structured energy determine **atomic stability and interactions**?
- Can an energy-based atomic model **explain chemical bonding better than quantum mechanics?**
- What experimental evidence would confirm that **particles are just energy configurations**?

Suggested Readings & References

- **The Failure of the Bohr Model** – Why classical atomic models misrepresent reality.
- **Energy-Based Alternatives to Quantum Mechanics** – How structured energy could replace wave-particle duality.
- **Chavan's Energy Based Atomic Configuration** – A new approach to atomic structure based on energy formations.

11: Gravity as an Energy Gradient

No Space-Time Needed

For over a century, gravity has been defined by the concept of space-time curvature. Newton described it as an invisible force acting across empty space. Einstein replaced this with the idea that mass bends space-time, influencing the motion of objects. Physicists accepted relativity because it provided accurate predictions—without ever questioning whether space-time was real.

But in Chapter 10, we uncovered a deeper reality. Quantum mechanics is not governed by randomness, but by structured vacuum energy. Particles are not discrete objects, but stable energy formations within the vacuum field. Forces are not independent interactions, but natural effects of energy density variations.

Now, we apply this same principle to gravity. What if gravity is not caused by the bending of space-time, but by energy density variations in the structured vacuum? What if mass does not warp "empty space," but simply alters the distribution of energy? What if light bends near massive objects, not because of curved space-time, but because of gradients in vacuum energy?

This chapter dismantles the illusion of space-time and redefines gravity as an energy-based phenomenon. Mass creates variations in energy density—not distortions in space-time. Gravitational lensing occurs not because of curvature, but due to shifts in energy gradients. Black holes are not singularities or "holes" in space but energy regulators within the vacuum field.

For too long, physics has been forcing gravity to fit a flawed framework. No more curved space-time. No more singularities. No more trying to reconcile relativity with quantum mechanics when the real solution is much simpler.

Gravity is not a distortion of geometry—it is an effect of structured energy. This is the final break from space-time.

11.1 The Misinterpretation of Gravity – How Physics Mistook Energy Gradients for Curved Space

For centuries, physicists have struggled to define the true nature of gravity. Newton described it as an invisible force acting across empty space. Einstein rejected this "action at a distance" and replaced it with the idea that mass bends space-time, influencing motion. General relativity became the accepted model, treating gravity as a geometric distortion rather than a force.

But what if gravity is neither a force nor a bending of space-time? What if it is simply an effect of energy density variations? What if mass does not warp an abstract geometry but instead alters the distribution of vacuum energy? What if light bends, not because space itself curves, but because it follows natural energy gradients? What if space-time is just a mathematical tool, not a physical entity?

This section exposes the flaws in the space-time model and reveals why gravity is better understood as an effect of structured vacuum energy.

Why Einstein Had to Abandon Newton's Gravity But Couldn't Fully Replace It

Newton's gravity was highly successful—but incomplete. It correctly predicted planetary motion, tides, and falling objects. It treated gravity as an instantaneous force acting across empty space. But it never explained what gravity actually was—only how it behaved.

Einstein saw the problem. How could the Sun "pull" on the Earth across millions of kilometers of empty space? How could gravity act instantly without a medium for transmission? Why did Newton's model fail at high speeds and strong gravitational fields? To solve this, Einstein proposed an alternative. Instead of treating gravity as a force, he suggested that mass bends space-time itself. Objects follow curved paths, not because they are "pulled," but because space itself is distorted. This removed the need for "instant action at a distance"—but introduced a new problem.

Einstein's space-time model was a fix—not a fundamental explanation. It replaced one mystery with another: What is space-time? It described how mass affects motion—but not why mass interacts with space at all. It ignored the possibility that space itself is an active, structured energy field.

Einstein's relativity was an improvement—but it was never a complete solution. He removed Newton's force but replaced it with an abstract curvature. He could not explain why mass bends space-time—only that it does. He could not unify gravity with quantum mechanics, which treats space as an energy field.

This is where modern physics made its biggest mistake. It treated space-time as a real physical thing—when it was just a mathematical framework. It ignored the role of structured vacuum energy—forcing gravity into a geometric model. It created the illusion that space itself "curves"—when in reality, energy density is changing.

Once we remove the artificial idea of space-time curvature, gravity becomes much easier to understand. Mass influences energy density, not geometry. Gravity emerges from structured energy gradients, not from bending nothingness. The entire framework of general relativity becomes unnecessary.

How Space-Time Curvature Was a Mathematical Fix, Not a Physical Explanation

General relativity is often praised as one of the greatest theories in physics. It makes accurate predictions about planetary motion, black holes, and gravitational lensing. It describes gravity as a smooth curvature of space-time.

It eliminated Newton's "force at a distance"—replacing it with a continuous fabric of reality.

But there is a problem. No one has ever observed space-time itself. No experiment has ever confirmed that space is a real physical entity that bends. Physicists assume it exists—because the equations work, not because there is direct evidence.

Einstein's theory is a mathematical success—but a physical illusion. It describes gravity well, but it does not explain the cause. It forces space to bend without defining what space actually is. It treats space-time as a fabric, yet it has no material properties.

If space-time is not a physical thing, then what is really happening? Gravity must be an effect of something real—something we can measure. Energy density variations provide a much better explanation than curved geometry. By treating space as structured vacuum energy, we replace space-time with a physical mechanism.

The question is simple: Would you rather believe that "empty space bends"—or that gravity is caused by measurable energy gradients? The answer is obvious. Space-time curvature is a convenient fix, not a fundamental reality.

Why Mass Influences Energy Density, Not Geometry

If space is not a fabric, then what does mass actually do? Mass does not bend geometry—it alters the surrounding energy field. It creates a gradient in vacuum energy, causing objects to follow paths of least resistance. Gravity is an effect of this energy distribution—not a distortion of space itself.

Think of a river flowing around a rock. The water's energy density changes near the rock, creating currents. Objects in the river follow the energy flow—not because the water is "curved," but because of density differences. This is exactly how gravity works.

The presence of mass alters energy density in the vacuum field. Objects move along these energy gradients—not because space is bending, but because energy is distributed unevenly. This explains why gravity behaves like a force—even though it is not one.

This also solves the biggest conflict in modern physics. Quantum mechanics treats space as a field full of energy fluctuations. General relativity treats space as an empty fabric that bends. But if gravity is an energy gradient, these two views become the same. No more contradictions. No need for quantum gravity theories. Just a single, unified framework based on structured energy.

The Fundamental Flaw of Bending "Nothing" – Why Space-Time Is Not a Physical Entity

The final and biggest mistake of modern physics is the belief that "nothing" can bend. Space-time is not a material substance. It has no measurable properties—other than what we assign to it mathematically. Yet physics treats it as if it has the ability to stretch, warp, and curve.

How can something with no physical existence bend? How can space have "shape" if it has no substance? If space is truly empty, what exactly is being curved?

The reality is: Space is not bending—energy is simply being redistributed. Gravity is not a warping effect—it is an energy interaction. Objects move through energy gradients, not curved nothingness.

By abandoning the illusion of space-time, we unlock the real physics of gravity. No more abstract geometric distortions. No more singularities or paradoxes. Just structured vacuum energy shaping the motion of matter.

Gravity is not a bending of space—it is an effect of energy. This is the truth physics has been avoiding for over a century.

Next: How Mass Creates Energy Density Variations, Not Space-Time Curvature

Why structured vacuum energy naturally explains gravity. How mass interacts with the vacuum field to create energy gradients. Why gravity behaves like an energy flow rather than a geometric warping. How this model resolves the conflict between quantum mechanics and relativity.

Because the truth is simple: Gravity is not geometry—it is energy.

11.2 How Mass Creates Energy Density Variations, Not Space-Time Curvature

For over a century, physics has been trapped in a false debate about the nature of gravity. General relativity treats gravity as a bending of space-time, ignoring the energy structure of space. Quantum mechanics, on the other hand, treats space as a fluctuating energy field but does not explain how this field interacts with mass. These two models have remained fundamentally incompatible—not because gravity is unsolvable, but because both describe space in conflicting ways.

But what if gravity is neither a force nor a curvature? What if gravity is simply an energy flow caused by variations in vacuum energy density? What if mass does not "bend space" but instead creates a gradient in structured energy? What if gravity behaves as an effect of energy distribution rather than a distortion of geometry?

This section explains why structured vacuum energy naturally produces gravity—without the need for space-time curvature.

Why Structured Vacuum Energy Naturally Explains Gravity

Physics has long assumed that space is empty. General relativity treats it as a mathematical background with no inherent properties. Quantum mechanics, however, contradicts this assumption by showing that space is full of energy fluctuations. Yet mainstream physics refuses to acknowledge the obvious—that space is an active, structured energy field.

If space is filled with structured vacuum energy, then gravity is simply an energy distribution effect. Mass does not "bend" an empty void—it alters the surrounding energy density. Objects move, not because of curved space-time, but because they follow natural energy gradients. Gravity is not a separate force—it is a property of the energy structure of space itself.

This perspective resolves one of the biggest contradictions in physics. Quantum mechanics and relativity have always treated space differently. But if space is structured energy, then both theories describe the same underlying reality. This naturally unifies quantum mechanics and gravity—without the need for exotic new theories.

How Mass Interacts with the Vacuum Field to Create Energy Gradients

If space is an active energy field, then mass does not simply "sit" in space—it interacts with it. A massive object disturbs the surrounding energy distribution. This disturbance creates a gradient in vacuum energy density. Objects move in response to this gradient—not because of a force, but because of an energy imbalance.

Think of a river flowing around a large rock. The rock does not "pull" the water—it simply alters the flow around it. Objects in the river follow the natural currents created by the disturbance. The water is not bending—the energy distribution is changing.

Gravity works in exactly the same way. Mass creates a low-energy zone in the vacuum field. Surrounding objects naturally move toward this lower-energy region. No force is needed—just an interaction between mass and energy density.

This explains several key observations in physics. All objects fall at the same rate in a gravitational field. Gravitational effects decrease with distance, following an inverse-square law. Gravity appears "instantaneous" in Newtonian mechanics because energy gradients are naturally present.

Gravity is not a distortion of space—it is a structured energy effect.

Why Gravity Behaves Like an Energy Flow Rather Than a Geometric Warping

General relativity treats gravity as a warping of space-time. But space-time is an abstract mathematical concept, not a physical entity. If space-time does not exist as a medium, then it cannot bend. Yet objects still fall, orbit, and experience gravitational effects.

So what is really happening? Gravity is not a bending of geometry—it is an energy flow. Objects move along energy gradients in the structured vacuum field. Their motion is guided by natural energy density variations, not "curved space."

Think of air pressure in the atmosphere. A low-pressure zone does not "pull" air toward it—it simply creates a region of lower energy. Air naturally flows into this region, following energy gradients. Gravity works in the same way—mass creates an energy variation, and objects move accordingly.

This explains why gravity is universal—because all mass interacts with vacuum energy. It explains why gravity is not quantized—because it is not a force, but a continuous energy interaction. It explains why gravity has no known particle carrier (graviton)—because it is not an exchange force.

Gravity is an effect of structured energy, not a force mediated by particle interactions. No need for gravitons. No need for space-time curvature. Just a simple, physical explanation based on energy gradients.

How This Model Resolves the Conflict Between Quantum Mechanics and Relativity

Physics has been stuck for decades trying to merge relativity with quantum mechanics. Relativity treats space as a flexible fabric, while quantum mechanics treats it as an energy field. Gravity is modeled as a geometric effect in relativity, but as a missing force in quantum mechanics. Attempts to quantize gravity have failed—not because gravity is too complex, but because gravity is not a force at all.

By recognizing gravity as an energy gradient, this conflict disappears. Space is not an empty void—it is structured vacuum energy. Mass does not bend geometry—it alters energy density. Gravity is not a force—it is a response to an energy imbalance.

This model naturally explains both quantum and relativistic effects. Quantum mechanics works because it describes local energy interactions. Gravity emerges as a large-scale effect of structured vacuum energy. The two models describe different scales of the same fundamental reality.

This also explains why current attempts to unify physics have failed. Physicists are trying to force gravity into a quantum force model. But gravity is not a force—it is an effect of energy distribution. This is why gravitons have never been detected—because they do not exist.

By shifting our perspective from space-time to structured energy, we finally see gravity for what it truly is. It is an emergent property of vacuum energy density variations. It is a natural result of energy flow, not a fundamental force. It is the bridge between quantum mechanics and large-scale physics—without contradictions.

This is the key to unifying physics. Gravity is not a warping of nothing—it is an energy effect. The vacuum is not empty—it is the foundation of all interactions. Mass does not "bend" space—it interacts with structured energy, shaping motion naturally.

This is the real nature of gravity—hidden behind the illusion of space-time curvature.

Next: Why Gravitational Lensing Can Be Explained Without Relativity

The traditional explanation: How relativity claims light follows curved space-time. The alternative: How energy density variations create a refractive effect. Why light bends in gravitational fields—without needing space-time distortion. How this model aligns with experimental observations while eliminating relativity's assumptions.

Because the truth is simple: Gravity is not a bending of space—it is an energy effect.

11.3 Why Gravitational Lensing Can Be Explained Without Relativity

For decades, physicists have attributed the bending of light around massive objects to space-time curvature. Einstein's general relativity predicts that gravity "warps" space, forcing light to follow curved paths. Observations of light bending around the Sun during solar eclipses confirmed this prediction, leading physicists to interpret this as proof that space itself is curved. Instead of questioning whether another mechanism was at play, the idea of space-time curvature became an unquestioned truth.

But what if gravitational lensing is not caused by space-time curvature at all? What if light bends due to energy density variations in the structured vacuum? What if the effect is better explained through energy field interactions rather than geometry? What if space is not bending—only the properties of light are changing as it moves through an altered energy landscape?

This section exposes the flaw in relativity's explanation and reveals how gravitational lensing can be understood without curved space-time.

The Traditional Explanation: How Relativity Claims Light Follows Curved Space-Time

Einstein's theory of general relativity replaced Newton's concept of gravity as a force with the idea of space-time curvature. According to relativity, mass bends the fabric of space-time. Objects—including light—follow the curved geometry created by massive bodies. This leads to the prediction that light passing near a star or galaxy will follow a curved trajectory.

This prediction was confirmed in 1919 during a solar eclipse when astronomers observed that starlight passing near the Sun appeared slightly displaced from its expected position. The bending matched Einstein's calculations, reinforcing the space-time curvature model. Physicists took this as confirmation that light follows the "warped" contours of space.

Since then, gravitational lensing has been treated as direct proof of general relativity. Observations of light bending around galaxies and black holes are cited as evidence that space-time itself is curved. Massive objects are believed to act as "lenses," bending and magnifying light from distant sources. The entire field of relativistic astrophysics is built upon this interpretation.

But there is a fundamental problem with this model. It assumes that space itself is a flexible fabric that bends—yet space has no measurable physical properties. It treats light as following a curved path—without explaining how this curvature physically occurs. It ignores the fact that bending light does not require geometry—only a change in the medium through which it travels.

What if gravitational lensing can be explained without bending space at all? What if light simply changes direction due to variations in vacuum energy density? This is where the real explanation begins.

The Alternative: How Energy Density Variations Create a Refractive Effect

Light does not need a curved space-time to bend—it only needs a medium with varying energy density. In optics, light bends when it moves between materials of different refractive indices. This is why a straw in a glass of water appears bent—because light changes speed as it moves through different densities. This effect is well understood and does not require bending space.

Now, consider what happens when light moves through a gravitational field. If space is structured vacuum energy, then mass alters the local energy density. A massive object, like the Sun, creates a gradient in vacuum energy. Light passing through this region experiences a change in energy conditions—just like light passing through glass or water. The result? Light bends—not because space is curved, but because its energy medium changes.

This model has key advantages over general relativity. It does not require space-time to be a physical entity. It explains light bending using known physical principles—energy density variations. It aligns with quantum mechanics, which already describes vacuum fluctuations.

If light bends due to energy field interactions rather than space-time warping, then gravitational lensing is not proof of relativity—it is proof that mass alters vacuum energy. Massive objects do not distort geometry—they create structured energy gradients. Light follows energy flow, just like any other wave in a medium.

This model eliminates the need for space-time curvature entirely. No more abstract bending of an undefined "fabric." No more contradictions between quantum mechanics and relativity. Just a simple, energy-based explanation of how gravity affects light.

Why Light Bends in Gravitational Fields—Without Needing Space-Time Distortion

In relativity, light bends because space is distorted. But this assumes that space is a "thing" that can bend. Since space itself has no material properties, this explanation is purely mathematical. It describes the effect, but it does not explain the cause.

In an energy-based model, light bends because mass creates a gradient in vacuum energy. This gradient alters the local properties of light's propagation medium. Light naturally follows the path of least resistance, adjusting its trajectory.

This is not warping—it is refraction. In air, sound waves change direction when they pass through temperature gradients. In water, light bends when it moves between different densities. In space, light follows energy variations created by mass.

This also explains why light behaves normally when far from massive objects—because energy density is uniform. It explains why gravitational lensing matches optical lensing—because both rely on medium variations. It explains why light speeds up or slows down near strong gravitational fields—because energy density affects propagation speed.

With this model, we no longer need to imagine space "bending" to explain light's motion. We remove unnecessary abstraction and replace it with known physical principles. We finally understand gravitational lensing as an optical effect—not a geometric distortion.

How This Model Aligns with Experimental Observations While Eliminating Relativity's Assumptions

Physicists claim that gravitational lensing confirms relativity. But the observations themselves only show that light bends. They do not prove that space is curved—only that light changes direction near mass. The explanation for this bending has been assumed, not proven.

By recognizing gravitational lensing as a refractive effect caused by energy gradients, we achieve a physical mechanism for why light bends—vacuum energy variations. We establish a natural link between quantum mechanics and gravity—both describe energy interactions. We provide a simpler explanation than relativity—without bending an undefined space-time fabric.

This model also works for all observed lensing effects—without requiring curved space. It explains why light behaves predictably in gravitational fields—just like waves in any medium. It replaces a mathematical abstraction (space-time curvature) with a real physical cause (energy density gradients).

By discarding relativity's geometric interpretation, we finally understand gravitational lensing correctly. Light does not follow a curved space-time—it moves through an energy field shaped by mass. Gravity is not a warping of space—it is a structured interaction of energy. Lensing is not proof of relativity—it is proof that space is an active energy medium.

This is the breakthrough physics has been avoiding. No space-time. No bending fabric. Just structured energy and natural wave behavior. This is how we move beyond Einstein's outdated model.

Next: The True Role of Black Holes as Energy Regulators, Not Singularities

Why black holes do not contain singularities—just extreme energy density regions. How mass accumulation alters vacuum energy structure without collapsing space. Why black holes are not "rips" in space-time, but dynamic energy balance points. How viewing black holes as energy regulators solves major inconsistencies in modern physics.

Because the truth is simple: Light bends because of energy gradients—not because space is curving.

11.4 The True Role of Black Holes as Energy Regulators, Not Singularities

For decades, black holes have been described as the ultimate breakdown of physics. General relativity predicts that massive stars collapse into a "singularity"—a point of infinite density and zero volume. Inside this singularity, space-time is said to "rip" apart, and the laws of physics break down. The event horizon is treated as an invisible boundary beyond which nothing escapes—not even light.

But what if black holes are not singularities at all? What if they are regions of extreme energy density—without any infinite collapse? What if mass accumulation alters vacuum energy, creating a stable but dense energy structure? What if black holes are not "holes" in space-time but dynamic energy regulators?

This section redefines black holes—not as mysterious singularities, but as natural stabilizers in the structured vacuum energy framework.

Why Black Holes Do Not Contain Singularities, Just Extreme Energy Density Regions

Physicists have long accepted the idea of singularities, despite their contradictions. A singularity is a mathematical construct where density becomes infinite. This means all mass is squeezed into an infinitely small point. Yet, infinite density is physically impossible—it is a sign that the model is incomplete.

Reality does not allow infinite quantities. Nature always finds a way to stabilize energy imbalances. Extreme conditions do not create singularities—they create new states of matter and energy. A better explanation is that black holes reach a critical density, forming a stable, ultra-dense energy state.

If we abandon the singularity myth and treat black holes as high-energy density objects, we resolve many paradoxes. No more infinite gravity at the center. No more physics breakdown—only extreme energy interactions. No need for space-time to "rip"—just a high-density energy configuration.

Black holes are not singularities—they are stable formations in structured vacuum energy. Just like an electron is a stable energy configuration at small scales, a black hole is a stable energy structure at large scales.

How Mass Accumulation Alters Vacuum Energy Structure Without Collapsing Space

General relativity assumes that when mass accumulates beyond a certain limit, space collapses. This leads to the prediction of an event horizon—a boundary beyond which gravity is so strong that nothing escapes. However, this assumes that space is an empty, flexible fabric. But we now know that space is structured vacuum energy—not a void.

If space is a structured energy field, then mass does not collapse it—it simply alters its density. A black hole is not an object collapsing into nothingness. It is a highly compressed energy zone within the structured vacuum. Instead of infinite density, it reaches a critical energy balance where forces stabilize.

Think of a neutron star. A neutron star is an ultra-dense object, but it does not collapse into a singularity. It stabilizes at a high-energy state where neutron degeneracy pressure prevents further collapse. Similarly, a black hole reaches an even higher density but remains a structured energy object.

This model explains why black holes do not destroy space—they modify vacuum energy density. Event horizons are not "points of no return"—they are transition zones of high energy gradients. Inside a black hole, mass-energy interactions continue—rather than "disappearing" into an unknown singularity.

If we view black holes as high-energy states rather than collapses, we eliminate the paradoxes of singularities. No more infinite density. No more violations of physics. Just a new form of stable energy configuration.

Why Black Holes Are Not "Rips" in Space-Time, But Dynamic Energy Balance Points

Traditional physics treats black holes as distortions in space-time. Space-time is said to "stretch" infinitely at the event horizon. Inside the black hole, time is theorized to stop. The singularity is viewed as a point where the fabric of the universe "tears."

But these ideas are based on the flawed assumption that space-time is a real entity. If space-time is just a mathematical tool, then it cannot bend or rip. If space is actually structured vacuum energy, then black holes are simply stable high-energy regions. Nothing "breaks" inside a black hole—only energy density increases beyond normal limits.

Black holes are not rips in reality—they are energy-dense zones where structured vacuum interactions dominate. Gravity inside a black hole does not become infinite—it stabilizes at an ultra-high energy state. Instead of being a "hole," a black hole is an ultra-stable configuration of vacuum energy.

Think of a whirlpool in water. It looks like a "hole," but it is actually a structured energy flow. Matter entering it does not disappear—it gets restructured within the system. Black holes function the same way—highly dense but stable energy configurations.

This model fixes the contradictions of black hole physics. No singularities. No broken laws of physics. Just a naturally occurring, high-energy vacuum structure.

How Viewing Black Holes as Energy Regulators Solves Major Inconsistencies in Modern Physics

The current black hole model creates more problems than it solves. Singularities violate known physics. Event horizons introduce paradoxes like information loss. Space-time curvature cannot explain what happens inside a black hole.

By replacing the space-time model with an energy-based model, we solve these issues. Black holes regulate excess energy by stabilizing it in a high-density state. They do not "destroy" information—it is stored in energy configurations. They do not break the laws of physics—they follow natural vacuum energy principles.

This model also explains why black holes emit jets of energy—because they are not dead zones, but dynamic systems. It explains why Hawking radiation works—because high-energy interactions at the event horizon create particle emissions. It explains why black holes do not violate thermodynamics—because they are stable, not infinite-density objects.

In this model, black holes are not mysterious anomalies. They are energy stabilizers in the structured vacuum. They regulate mass-energy interactions in high-density zones. They play a crucial role in balancing the universe's energy distribution.

This is the key to truly understanding black holes. No more singularities. No more broken physics. Just structured energy dynamics in action.

This is how modern physics must evolve.

Key Takeaways

- Gravity is **not a force and not a curvature of space-time**, but an effect of **energy gradients in structured space.**
- Einstein's relativity **redefined gravity as space-time curvature**, but this is a **mathematical interpretation, not a physical mechanism.**
- The **existence of gravity without force carriers** (gravitons) suggests it is an emergent phenomenon rather than a fundamental force.
- Mass does not "bend space"—instead, it **interacts with the structured energy field of space**, creating an energy density gradient that produces gravitational effects.
- The need for **dark matter to explain missing mass in galaxies** suggests that physics is missing a crucial understanding of how gravity emerges.
- An **energy-based gravity model** naturally unifies with electromagnetism and quantum mechanics, something relativity fails to achieve.

What If...? (Speculative Exploration)

- What if gravity **is just a pressure effect of energy density gradients in space**, rather than a separate force?
- What if mass-energy **alters the structure of space**, influencing motion without bending an imaginary space-time fabric?
- What if dark matter **is not real**, but instead a misinterpretation of structured energy interactions?
- What if an **energy-based approach to gravity** naturally explains phenomena like black holes, galactic rotation, and cosmic expansion?

Why Has Physics Clung to Space-Time Curvature?

- General relativity **worked mathematically**, so physicists did not look for an alternative explanation.
- Gravity is a unique force because it **affects all objects equally**, which led to space-time being considered the "fabric" that carries it.
- The success of relativity in **predicting planetary motion and gravitational lensing** made it difficult to challenge.
- Physics lacks a quantum theory of gravity, but instead of questioning space-time, it keeps adding layers of complexity.
- The idea of **structured vacuum energy** is ignored because it contradicts the mainstream approach to fundamental forces.

Challenges & Open Questions

- If gravity is an energy gradient, **how do we measure the structure of space?**
- Can an **energy-based model of gravity** replace the need for dark matter?
- If gravity emerges from structured energy, **how does it connect with electromagnetism?**
- Could an alternative theory of gravity **lead to a true unified physics model**?

Suggested Readings & References

- **The Failure of Graviton Theory** – Why gravity has no force carrier.
- **The Contradictions in General Relativity** – Why Einstein's model is an approximation, not a fundamental truth.
- **Chavan's Grand Unified Theory of Physics** – A deeper exploration of gravity as an energy gradient.

12: Redefining Light

No Photons, Just Energy Flow

For over a century, physics has struggled with the fundamental nature of light. Sometimes, light behaves like a wave, spreading, diffracting, and interfering with itself. Other times, it behaves like a particle, appearing to transfer energy in discrete packets called "photons." Instead of resolving this contradiction, physicists accepted an artificial fix—wave-particle duality.

But in Chapter 11, we saw how gravity was misunderstood as a bending of space-time, when in reality, it is simply an effect of energy density variations in structured vacuum energy. If space is structured energy, then all physical interactions—including light—must be reconsidered in this framework. If gravity is not a force but an energy effect, then light is not a particle but an energy flow. If mass affects energy density, then light must be a disturbance in the same structured medium.

What if light has never been a particle at all? What if the photon is just another artificial "fix" invented to explain incomplete theories? What if light is simply an energy transmission through structured vacuum, with no need for discrete particles? What if electromagnetic waves are just fluctuations in the vacuum energy field, without the need for force-carrying quanta?

This chapter redefines the very nature of light—not as a stream of particles, but as a structured energy flow in vacuum energy. No more photons. No more wave-particle duality. Just a continuous, natural propagation of energy through structured space.

This is the final break from outdated quantum interpretations.

12.1 The Particle Illusion – Why Light Is Not Made of Photons

For over a century, physics has struggled with an inherent contradiction in its understanding of light. It behaves like a wave, spreading, diffracting, and interfering with itself. Yet, in other cases, it appears to transfer energy in discrete packets, which physicists call "photons." Instead of resolving this contradiction, science accepted an artificial fix—wave-particle duality.

The photon model was never a discovery—it was a compromise. It was introduced not because light was proven to be a particle, but because quantum mechanics needed a way to explain energy quantization. Light behaves like a wave in every experiment—except when forced into a particle-based interpretation. But what if the photon is just another artificial construct? What if light has never been a particle but has always been a structured energy flow? What if energy quantization is not due to "photon particles" but due to interactions between light and structured vacuum energy?

This section dismantles the photon myth and redefines light as a continuous wave of energy propagating through structured vacuum.

How the Photon Model Was a Convenient Fix, Not a Fundamental Truth

Before Einstein's explanation of the photoelectric effect in 1905, light was universally accepted as a wave. Maxwell's equations described it beautifully, and experiments confirmed interference, diffraction, and polarization—phenomena that only waves exhibit. No one had ever observed a "photon"—only energy exchanges that were interpreted as particle-like behavior.

So why did Einstein introduce the photon concept? Classical physics struggled to explain why light ejected electrons from metals in the photoelectric effect. Instead of questioning the deeper nature of light-matter interactions, physicists created an artificial solution: photons. The idea was that light came in "packets" of energy, solving the quantization problem—but at the cost of breaking wave theory.

This was not a discovery—it was a convenient fix. No one had ever directly observed an individual photon. No one had proven that light was fundamentally discrete. But the equations worked, and physics was more focused on solving immediate problems than questioning assumptions.

The photon quickly became accepted as reality, despite its contradictions. It explained energy quantization but created deeper paradoxes. It forced light to be both a particle and a wave—without explaining how. It introduced unnecessary complexity where a deeper energy-based explanation was possible.

Why Treating Light as Particles Contradicts Its Observed Behavior

If light were truly made of particles, it should behave like one in all experiments. Yet, it does not. Photons should scatter like classical particles, but they interfere like waves. Photons should follow definite paths, yet they exhibit diffraction and self-interference. Photons should have clear, localized positions, yet their energy appears distributed across space.

In reality, every major experiment confirms light's wave nature. Young's double-slit experiment proves that light travels as a wave, not as individual particles. Diffraction patterns appear even when light is sent one "photon" at a time—showing that the wave exists even when we try to isolate particles. Light polarization cannot be explained by particle interactions—only by wave behavior.

The idea of photon particles contradicts these observations. If photons were real particles, they should not interfere with themselves. If photons were real particles, their behavior should not depend on observation. If photons were real particles, we should be able to isolate them in space—yet we cannot.

In reality, all experiments that "prove" photons exist rely on indirect interpretations. We detect energy exchanges—not individual particles. We measure discrete absorption events—but that does not prove that light traveled as a particle. We force wave behavior into a particle-based explanation—rather than questioning the assumptions.

By insisting that photons exist, physics ignores the more natural explanation: Light is not a particle—it is a wave propagating through structured vacuum energy.

The Real Nature of Light: An Energy Wave Propagating Through Structured Vacuum

If space is structured vacuum energy, then light is not a particle at all. Light is a self-propagating wave in an energy field, just as Maxwell originally described. It moves through space because space itself is not empty—it is an energy medium. It interacts with matter in quantized ways because matter itself is an energy structure.

This explains why light interferes with itself—because it is an extended wave, not a discrete object. It explains why the speed of light is constant—because it depends on the properties of the structured vacuum, not on the motion of an observer. It explains why light does not require a particle model—because energy waves naturally exhibit quantization without needing photons.

Think of a ripple on the surface of a pond. It carries energy across the water, but no "ripple particle" exists. The energy propagates through the medium itself, not as discrete objects moving through space. The ripple is an effect of the water's structure—just as light is an effect of vacuum energy structure.

When we remove the artificial photon model, light's behavior becomes simple and intuitive. It is a wave in an energy medium. It propagates naturally, without particles or force carriers. It interacts with matter through structured energy exchanges—not by "photon impacts."

This model resolves all major contradictions in modern physics. No more paradoxes of wave-particle duality. No more unexplained behavior in double-slit and diffraction experiments. No more unnecessary quantization assumptions—only natural energy interactions.

Light has never been a particle. It is an energy flow—continuous, structured, and wave-like.

How Abandoning the Photon Model Resolves Contradictions in Quantum Mechanics

The biggest problem in quantum mechanics is the artificial separation of waves and particles. Matter is treated as particles, but with wave-like properties. Light is treated as waves, but also as discrete particles. Wave-particle duality is accepted as "true"—despite the contradictions.

But if we abandon the photon model, all of these problems disappear. Light is purely a wave—so no need to force it into a particle interpretation. Electrons and other "particles" are actually stable energy structures—not point-like objects. Energy quantization is not due to "photon absorption" but due to interactions within the structured vacuum.

This fixes the confusion around wave-particle duality by removing the need for particles. It resolves inconsistencies between relativity and quantum mechanics by treating both as structured energy theories. It explains why we fail to detect individual photons—because they do not exist as real objects.

By redefining light as a structured wave in vacuum energy, we move beyond outdated physics and into a true unified model. No photons. No contradictions. Just energy waves propagating through a structured medium.

This is the reality of light—hidden for over a century behind artificial fixes.

Next: The Photoelectric Effect Revisited – How It Works Without Photons

The photoelectric effect was Einstein's key argument for photons as "quanta of light," but was it necessary? The next section reexamines this phenomenon, showing that electron ejection is actually an interaction between energy waves and structured vacuum fields. Energy transfer is not about particle collisions—it is about resonance and wave interactions.

Understanding this not only explains the photoelectric effect without photons, but it also unifies quantum mechanics and classical wave theory. Because the truth is simple:

Light is not a particle—it is a structured energy wave.

12.2 The Photoelectric Effect Revisited – How It Works Without Photons

For over a century, the photoelectric effect has been presented as the strongest evidence that light is made of photons. Einstein's 1905 explanation introduced the idea that light consists of discrete packets of energy—later called photons. This concept was necessary to explain why light seemed to eject electrons from metal surfaces only above a certain frequency. Since then, physicists have taken this as proof that light must be both a wave and a particle.

But what if this interpretation was flawed from the beginning? What if the photoelectric effect is actually an energy transfer phenomenon, not a particle impact? What if light is a wave interacting with a structured vacuum energy field, rather than a stream of photons? What if electron ejection is a resonance effect—not a discrete quantum event?

This section revisits the photoelectric effect and shows how it can be fully explained without the need for photons.

Why Einstein Introduced Photons as "Quanta of Light"—and Why It Was Unnecessary

At the time of Einstein's work, physics was struggling to explain why light's energy absorption seemed quantized. Classical wave theory predicted that increasing light intensity should continuously increase energy absorption.

Yet, experiments showed that below a certain frequency, no electrons were ejected—no matter how intense the light. This suggested that light transferred energy in discrete amounts, leading Einstein to propose that light itself must be composed of discrete quanta.

However, Einstein's explanation was based on an assumption, not direct proof. Just because energy was absorbed in discrete amounts did not mean light itself was made of particles. Wave interactions can also be quantized if the system itself has natural frequency constraints. Instead of questioning the energy structure of the electron field, physics assumed that light must be quantized at the source.

Einstein's photon model was accepted because it fit the data—but it was never the only possible explanation. The quantization of energy could come from the nature of the electron's interaction with the vacuum field. The need for photons disappears if we realize that energy absorption is a resonance effect, not a particle exchange. The entire problem vanishes when we understand light as a structured wave in an energy medium.

The photon model was a shortcut, not a fundamental discovery. It ignored the deeper nature of how light interacts with electrons. It treated the quantization of energy as proof of particle-like behavior—when it was actually proof of wave resonance. It forced light into a quantum model rather than reconsidering assumptions about energy transfer.

This leads to a better question: If the photoelectric effect is not proof of photons, what is really happening?

How the Photoelectric Effect Is Actually an Interaction Between Energy Waves and Structured Vacuum Fields

If space is structured vacuum energy, then electron interactions are not isolated events. Electrons do not "sit" in empty space—they exist in an energy field. Light does not "hit" electrons like a particle—it interacts with the vacuum energy structure. Electron ejection does not require discrete photon packets—it requires a resonance condition to be met.

Here's how the photoelectric effect actually works in an energy-based model:

Light is an electromagnetic wave propagating through structured vacuum energy. The electron exists in a quantized energy state due to its interaction with this vacuum field. When light waves of a certain frequency match the electron's resonance condition, the electron absorbs energy efficiently. If the resonance condition is strong enough, the electron breaks free from its bound state—ejecting from the metal.

This explains all key observations of the photoelectric effect—without needing photons. The threshold frequency exists because energy transfer depends on resonance, not intensity. Increasing intensity without increasing frequency does not eject electrons—because resonance has not been met. Energy is absorbed in discrete amounts because the electron's interaction with the vacuum energy field is quantized—not because light is made of particles.

This model shows that energy exchange is an interaction between waves—not a particle collision. No photons. No contradiction between wave theory and quantum mechanics. Just natural energy resonance in a structured vacuum field.

Why Electron Ejection Is a Result of Resonance, Not Particle Impact

One of the biggest problems with the photon model is that it assumes electrons "catch" photons like baseballs. This treats electrons like tiny solid spheres waiting to be "hit" by discrete energy packets. But electrons do not exist as localized particles—they are energy structures interacting with a field. Their motion, absorption, and ejection must be understood as energy waves coupling with other energy waves.

Resonance explains this process far more naturally. When a wave interacts with an object at its natural frequency, the energy transfer is maximized. If the incoming light wave resonates with the electron's natural frequency, energy is absorbed efficiently. If the energy reaches a critical level, the electron is freed—exactly what we observe in the photoelectric effect.

Think of a child on a swing. A gentle push at the right frequency can build up energy and make the swing go higher. No "particle impact" is required—just a properly timed energy transfer. This is exactly how light waves transfer energy to electrons in the photoelectric effect.

This explains why only certain frequencies eject electrons—because only certain resonance conditions exist. It explains why energy is quantized—not because light is a particle, but because electrons have quantized energy levels. It explains why intensity does not matter—because without the right frequency, no resonance occurs.

This removes the need for photons entirely. The electron is not absorbing "light particles"—it is interacting with energy waves in structured vacuum.

How This Interpretation Unifies Quantum Mechanics and Classical Wave Theory

The photon model created an unnecessary divide in physics. It forced light into a particle-based explanation that contradicted its wave behavior. It separated quantum mechanics from classical wave theory—treating them as incompatible. It introduced an artificial problem: Why does light sometimes behave as a wave and sometimes as a particle?

By recognizing light as a structured energy wave, we fix these contradictions. Light behaves like a wave because it is a wave. Energy absorption appears quantized because electrons have discrete resonance conditions—not because light is made of particles. The photoelectric effect is a wave interaction—not a particle collision.

This unifies quantum mechanics by showing that energy quantization arises from structured field interactions, not discrete particles. It unifies classical electromagnetism by confirming that Maxwell's wave theory is correct, without needing a quantum fix. It clarifies the nature of light by proving that photons were never needed in the first place.

The photoelectric effect was never proof of photons—it was proof that energy interactions follow wave resonance principles. Einstein's model worked mathematically, but it was based on the wrong assumptions. Light is not a particle—it is a structured energy wave in vacuum energy.

Next: Why Electromagnetic Waves Are Structured Energy Flows, Not Particle Streams

Einstein's interpretation of the photoelectric effect led to the concept of photons, but it was never necessary. The next section explores why Maxwell's equations already describe light as a wave—without needing particles. It examines how the vacuum is not empty but an active energy medium. It explains why electromagnetic waves propagate through vacuum energy fluctuations, following structured patterns.

Understanding this not only confirms that light behaves like a wave in all experiments, but it also eliminates the need for force-carrying quanta. Because the truth is simple:

Light is not a stream of photons—it is an energy wave interacting with structured vacuum.

12.3 Why Electromagnetic Waves Are Structured Energy Flows, Not Particle Streams

For over a century, physics has struggled with a contradiction at the heart of light's nature. Maxwell's equations describe light as a continuous wave. Quantum mechanics, however, introduced the idea of photons—discrete packets of light energy. To reconcile these two opposing views, physics invented wave-particle duality, an artificial fix that allowed light to be both a wave and a particle depending on how it was observed.

But what if this contradiction exists only because physics took a wrong turn? What if light is not a particle at all, but simply an energy wave moving through structured vacuum? What if photons are an unnecessary mathematical fix rather than a physical reality? What if electromagnetic waves propagate as natural oscillations in an energy medium—without the need for discrete particles?

This section redefines light and electromagnetic waves as structured energy flows, eliminating the need for photons entirely.

How Maxwell's Equations Already Describe Light as a Wave—Without Needing Particles

Long before quantum mechanics introduced photons, James Clerk Maxwell provided the most accurate description of light. His equations showed that electric and magnetic fields oscillate and propagate through space as waves. His theory predicted that these waves travel at a fixed speed—now known as the speed of light. It explained interference, diffraction, and polarization—phenomena that only waves can exhibit.

Yet, despite this elegant and successful theory, physics later introduced photons. Why? Because of the need to explain energy quantization in the photoelectric effect. Instead of questioning the nature of electron interactions, physics assumed light must be discrete. This assumption contradicted Maxwell's continuous wave description but was accepted anyway.

But Maxwell's equations never required particles. They describe how electromagnetic waves self-propagate through interactions between electric and magnetic fields. They show that light does not need a medium like air or water—because space itself is structured vacuum energy. They provide a complete mathematical framework for light's motion—without ever suggesting that light must be made of quanta.

If Maxwell's theory already describes light perfectly, why introduce photons? Because physics misunderstood the nature of energy absorption. Because quantized energy exchange does not mean that light itself is quantized. Because photons were a quick-fix solution, not a fundamental reality.

Once we remove the artificial concept of photons, Maxwell's equations remain the best description of light. Light is not a stream of particles—it is a structured, self-sustaining energy wave.

Why the Vacuum Is Not Empty but an Active Energy Medium

Physics has long treated space as "empty"—except when convenient. General relativity assumes space is a passive fabric that can bend, but otherwise has no properties. Quantum field theory contradicts this by stating that space is filled with fluctuating energy fields. Instead of resolving this contradiction, physics treats space as "nothing" in some cases and "something" in others.

But what if space is not empty at all? What if the vacuum is a structured energy medium that supports wave propagation? What if electromagnetic waves do not move through "empty space," but through a fundamental energy field? What if light is simply a wave disturbance in this structured vacuum—just like ripples on water?

This resolves one of the biggest mysteries of modern physics. How can light move through space if space is empty? Why does light always travel at a fixed speed, regardless of the observer? Why does light behave like a wave in every experiment—except when forced into a particle interpretation?

If space is a structured vacuum energy field, then light is not traveling through "nothing." It is propagating as a self-sustaining wave in an active medium. It does not need particles to carry energy—because the energy is stored in the field itself. It moves at a fixed speed because it interacts with the underlying energy structure of space.

This explains why the speed of light is always the same—because it is set by the properties of structured vacuum energy. It explains why light does not need a medium like air or water—because space itself is already an energy medium. It explains why electromagnetic waves exhibit classical wave behavior—because they are true waves, not particle streams.

The vacuum is not empty—it is the foundation of all physical interactions. And electromagnetic waves are simply structured oscillations in this fundamental energy field.

How Electromagnetic Waves Propagate Through Vacuum Energy Fluctuations

If space is structured vacuum energy, then light moves just like any other wave in a medium. Waves do not travel as discrete particles—they propagate as continuous disturbances in a field. Light is no different—it is an energy fluctuation moving through the structured vacuum. This means that light is not "emitted" as a photon—it is generated as a wave pulse that moves through space.

Think of how sound travels through air. Air molecules do not move from the speaker to your ear—only the wave does. The energy propagates as pressure variations in the air, not as a stream of particles. Similarly, light propagates as oscillating energy variations in structured vacuum.

This explains why electromagnetic waves do not require a medium in the traditional sense—because space itself is already an energy field. It explains why light waves maintain their structure over vast distances—because they are stable oscillations in vacuum energy. It explains why light does not behave like a stream of particles—because it is a continuous process, not a discrete event.

Light is not "traveling photons"—it is an oscillating energy flow in the structured vacuum field. No need for particles. No need for wave-particle duality. Just a simple, natural wave propagation mechanism.

Why Light Behaves Like a Wave in All Experiments—Except When Misinterpreted Through Particle Assumptions

The wave nature of light has been confirmed in every major experiment. Young's double-slit experiment shows that light exhibits interference—something only waves can do. Diffraction patterns confirm that light bends around obstacles, just like any classical wave. Polarization effects can only be explained by wave behavior, not particle impacts.

Yet, physics continues to insist that light is also a particle—despite the evidence. Why? Because of misinterpretations of energy quantization. The fact that light energy is absorbed in discrete amounts does not mean light itself is discrete. Wave interactions can also be quantized if the receiving system has quantized energy states.

When we remove the assumption that light must be a particle, all contradictions disappear. Light behaves like a wave because it is a wave. Energy quantization happens because of resonance effects, not because of particle exchanges. The double-slit experiment does not show wave-particle duality—it shows interference, proving that light is a wave.

By insisting on photons, physics has ignored the simplest explanation for over a century. Light is not a particle. It is a structured energy wave propagating through a vacuum energy field. This is the reality of electromagnetic waves—hidden beneath artificial quantum fixes.

Next: The Future of Light – Replacing Photons with a Structured Energy Model

Einstein's interpretation of the photoelectric effect led to the concept of photons, but it was never necessary. The next section explores why redefining light as structured energy flow solves paradoxes in physics. It examines why quantum optics must be rewritten to account for vacuum energy interactions. It explains how this model naturally accounts for interference, diffraction, and entanglement.

Understanding this not only clarifies the true nature of light, but it also opens new doors for future physics and technology. Because the truth is simple:

Light is not a photon stream—it is an energy wave moving through structured vacuum.

12.4 The Future of Light – Replacing Photons with a Structured Energy Model

For over a century, physics has been stuck in an unnecessary contradiction. Light is treated as both a wave and a particle—despite the fact that no one has ever observed a photon moving through space. Wave-particle duality was introduced not because it was discovered, but because it was needed to make quantum mechanics work. Instead of resolving the contradiction, physics chose to accept it as a fundamental aspect of reality.

Yet, every experiment confirms that light behaves as a wave. It diffracts, interferes, polarizes, and propagates exactly as Maxwell's equations describe. The only time it appears to behave like a particle is when energy is exchanged—yet that is an interaction effect, not proof of discrete photons. Despite this overwhelming evidence, mainstream physics refuses to abandon the photon model.

But now, we have reached a turning point. If light is a structured energy wave propagating through vacuum energy, then all contradictions disappear. If photons were never real, then we no longer need wave-particle duality. If electromagnetic waves are simply oscillations in structured vacuum energy, then light is fully explained without quantum paradoxes.

This section lays out the future of physics—where light is no longer misunderstood as a particle, but recognized as a structured energy flow.

How Redefining Light as Structured Energy Flow Solves Paradoxes in Physics

If light is not a photon stream but a structured energy wave, then many long-standing contradictions in physics are resolved. No more wave-particle duality—because light is just a wave. No more artificial quantization of light—because energy transfer is a result of resonance, not particle impact. No more unexplained speed of light invariance—because it emerges naturally from structured vacuum energy.

This model immediately solves several major paradoxes:

Why does light always move at a constant speed?

Because it is a wave propagating in structured vacuum energy, and its speed is determined by the properties of that medium.

Why do photons behave like waves in interference experiments?

Because photons do not exist—light is a wave, always.

Why does light behave like a wave in diffraction and polarization but a particle in energy transfer?

Because energy transfer is discrete due to the receiving system's properties, not because light itself is quantized.

By shifting to a structured energy model of light, physics eliminates contradictions instead of trying to work around them. No more photons. No more unexplained duality. Just a simple, unified wave model of electromagnetic energy.

Why Quantum Optics Must Be Rewritten to Account for Vacuum Energy Interactions

If light is a structured energy wave, then quantum optics must be rebuilt from the ground up. The current framework is built on the assumption that photons are real particles. It treats light as a stream of quanta rather than a continuous wave, and it builds entire models around particle behavior that has never actually been observed.

But if light is an energy wave interacting with vacuum fluctuations, then the entire foundation of quantum optics must change. Quantum optics should describe structured wave interactions—not photon collisions. Energy exchange must be understood as wave resonance effects—not particle absorption. Entanglement must be reinterpreted as energy coherence in the structured vacuum—not spooky action at a distance.

This means many existing quantum mechanical concepts must be revisited:

- **Quantum states** should be seen as energy configurations in vacuum fields—not probabilistic particle states.
- **Light-matter interaction** must be understood through structured vacuum interactions—not photon exchange.
- **Lasers, fluorescence, and absorption** must be modeled as energy wave resonance—not discrete photon events.

If physics accepts that space is not empty but an energy medium, quantum optics will finally align with reality. No more photon-based explanations. No more probability-based light interactions. Just structured waves interacting with structured vacuum energy.

How This Model Naturally Explains Interference, Diffraction, and Entanglement

One of the biggest failures of the photon model is its inability to naturally explain core optical phenomena. Interference experiments show that light can cancel itself out—something only waves can do. Diffraction patterns emerge because light bends around obstacles—just like any classical wave.

Polarization effects are based entirely on wave properties—not particle motion.

Yet, physicists continue to force photons into explanations of these effects. They claim photons "interfere with themselves"—a contradiction that makes no logical sense. They argue that diffraction happens because photons "probabilistically" travel different paths. They suggest that light polarization works because photons "align"—despite no evidence of this happening.

But if light is simply a structured energy wave, then all of these effects are naturally explained:

- **Interference** occurs because light waves combine constructively or destructively—no particles needed.
- **Diffraction** happens because light, as a wave, interacts with the edges of objects.
- **Polarization** is just a directional oscillation in structured vacuum energy—not a property of individual particles.

This model also provides a natural explanation for quantum entanglement. Entangled photons are not "spooky particles" communicating faster than light. They are two parts of the same coherent energy wave, connected through structured vacuum energy. When one is measured, the entire wave structure is affected—not because of "instant communication," but because it was never truly separate.

By recognizing light as an energy wave rather than a photon stream, physics finally has a coherent explanation for optical phenomena. No more mysterious self-interference. No more probabilistic diffraction. No more "spooky" entanglement. Just a structured wave system operating exactly as expected.

The Implications of an Energy-Based Light Theory for Future Physics and Technology

Once physics accepts that light is a structured energy wave, the impact on science and technology will be enormous. New optical technologies will be developed that manipulate vacuum energy fields directly. Quantum computing will shift from particle-based entanglement to coherent wave-based interactions. Energy transmission will be revolutionized by understanding light as a field effect rather than a stream of particles.

This model also opens the door for new discoveries in:

- **Vacuum energy engineering**—tapping into the structure of space itself.
- **Non-local communication**—harnessing energy coherence instead of information transfer.
- **Wave-based computing**—moving beyond digital electronics to continuous field processing.

And perhaps most importantly, a **complete unification of physics**, eliminating artificial quantum paradoxes.

By embracing light as a structured energy wave, we take the final step toward a new scientific revolution. No more photon confusion. No more incomplete quantum optics. Just a unified, wave-based understanding of reality.

This is the future of light—the way it was meant to be understood.

Key Takeaways

- Light is not made of **discrete photons** but is instead a **continuous energy flow within a structured medium.**
- The **wave-particle duality paradox** exists only because physics insists on treating light as both a wave and a particle instead of recognizing it as an energy transfer phenomenon.
- The **photoelectric effect does not prove that photons exist**—it only shows that energy is absorbed and released in discrete amounts, which can be explained through structured energy interactions.
- The assumption that light must be **either a wave or a particle** is flawed; instead, light is **a structured disturbance in the energy field of space.**
- The speed of light is **not a universal limit** but a property of the medium in which energy propagates.
- If light is simply a structured energy flow, then **forces, electromagnetism, and even gravity can be understood through the same energy principles.**

What If...? (Speculative Exploration)

- What if light **is not an object or a particle**, but simply a process of energy transfer in space?
- What if there are **no individual photons**, only structured waves in an underlying energy field?
- What if the speed of light **is not a fixed universal constant**, but a property that depends on the medium it moves through?
- What if removing the concept of photons **allows us to unify light, gravity, and electromagnetism into a single framework?**

Why Has Physics Clung to Photons?

- The **photoelectric effect** forced physics to introduce a particle-like concept for light, but this was a mathematical workaround, not a proof of physical existence.
- Quantum mechanics needed photons to explain discrete energy levels in atoms, **but this was just a convenient assumption.**
- Experiments like the **double-slit experiment** are misinterpreted as proof of wave-particle duality instead of structured energy interactions.
- The Standard Model treats **everything as particles**, even when better explanations exist.
- The **structured energy flow model** was ignored because it contradicts mainstream quantum mechanics.

Challenges & Open Questions

- If photons don't exist, **what exactly is happening in the photoelectric effect?**
- How does structured energy explain **polarization, diffraction, and interference?**
- Can an energy-based model **unify light, electromagnetism, and gravity?**
- If light is an energy flow, **what determines its speed and behavior in different media?**

Suggested Readings & References

- **The Misinterpretation of the Photoelectric Effect** – Why photons may not exist.
- **Light as an Energy Flow, Not a Particle** – Exploring structured energy theories.
- **Chavan's Grand Unified Theory of Physics** – A deeper explanation of light as a structured energy transfer.

13: The Future of Science

Breaking Free from the Fixes

For over a century, modern physics has been built on a foundation of quick fixes rather than fundamental truths. Instead of addressing contradictions at their core, physics has patched over them with artificial solutions. Every major "breakthrough" in quantum mechanics and relativity has been another workaround—an adjustment made to maintain an incomplete model rather than a step toward genuine understanding.

Despite decades of research, quantum mechanics and relativity remain incompatible. Instead of resolving their contradictions, physics has chosen to work around them, treating the conflict as an unsolvable mystery rather than an indication that something is fundamentally wrong. The Standard Model, once seen as a great achievement, has become a system of mathematical adjustments—adding new particles, forces, and interactions whenever the theory fails to match observations. It does not predict reality; it conforms to it after the fact.

But now, we stand at the edge of a new scientific revolution. We have seen that structured vacuum energy eliminates the need for space-time curvature, particle force carriers, and wave-particle duality. We have uncovered that light is not a photon stream but an energy wave propagating through structured space. We have dismantled the illusion of mathematical physics and restored reality to a foundation of physical principles.

If physics is to move forward, it must finally break free from the patchwork model. No more forced dualities. No more "discoveries" that serve as placeholders for missing theory. No more reliance on abstract equations detached from physical reality. Science must return to its original goal—not adjusting theories to fit incomplete models, but uncovering the true nature of the universe.

This chapter lays out the future of science—the path beyond temporary fixes, where physics is once again guided by fundamental understanding rather than mathematical improvisation.

13.1 The Patchwork Collapse – Why Physics Must Abandon Its Fixes

Modern physics was once a quest for truth—a pursuit to uncover the fundamental nature of reality. It aimed to describe the universe through elegant, universal principles, revealing the deeper order behind existence. Yet, over the past century, physics has drifted away from its original purpose, becoming a system of patches and quick fixes rather than fundamental understanding.

Whenever contradictions arose, physics responded not by reevaluating its assumptions, but by adding more layers of complexity. When Newton's gravity failed to explain Mercury's orbit, Einstein introduced space-time curvature. When classical physics couldn't explain the photoelectric effect, the photon was invented. When atomic structure defied classical models, determinism was abandoned in favor of quantum probability. And when relativity and quantum mechanics proved incompatible, physics sidestepped the problem rather than solving it.

With every contradiction, a new "fix" was introduced, yet never a true resolution. This is why physics remains fragmented today. General relativity describes gravity but has no connection to quantum mechanics. Quantum mechanics describes atomic behavior but cannot explain gravity. The Standard Model, instead of offering a unified picture, is a constantly expanding collection of particles and forces, each added to compensate for previous gaps.

Rather than seeking deeper principles to unify physics, more patches continue to appear. Dark matter was introduced when gravity failed on galactic scales. Dark energy was added when the universe's expansion didn't match predictions. Supersymmetry, string theory, and extra dimensions—all attempts to resolve contradictions—are not discoveries, but compensations for an incomplete framework.

This is not progress—it is intellectual survival. Physics no longer seeks truth; it merely attempts to prevent its own collapse.

Why Quantum Mechanics and Relativity Cannot Be Unified Under Current Models

For over a century, physicists have tried to reconcile quantum mechanics and relativity, yet the two theories remain fundamentally incompatible.

Relativity treats space and time as a continuous, bending fabric. Quantum mechanics, on the other hand, reduces everything to discrete energy interactions. Relativity describes gravity as a smooth curvature, while quantum mechanics insists that nature must be quantized. Relativity assumes that objects exist with definite properties, but quantum mechanics claims that particles remain in probabilistic states until measured.

These contradictions run deep. In relativity, time is relative to the observer; in quantum mechanics, time remains absolute in equations. Relativity treats space as an active participant, bending under mass, while quantum mechanics treats space as a passive stage where quantum events unfold. Relativity assumes that energy is continuous, while quantum mechanics requires energy to be exchanged in discrete packets.

Rather than recognizing that both theories are incomplete, physics has spent a century forcing them together. The result has been an endless stream of theoretical fixes. String theory adds extra dimensions to make equations work. Loop quantum gravity attempts to quantize space-time itself, ignoring the deeper issue of vacuum energy structure. The search for the graviton—the hypothetical quantum of gravity—continues despite gravity behaving nothing like other fundamental forces.

The real problem is that both theories are built on flawed foundations. Quantum mechanics assumes that particles are fundamental, when in reality, they are structured energy configurations. Relativity assumes that space-time is a real entity, when in fact, it is an emergent effect of energy interactions. Both fail to recognize that structured vacuum energy is the missing link—the bridge between the macroscopic and microscopic universe.

Physics cannot unify quantum mechanics and relativity because both are incomplete. The real solution is not to patch them together, but to replace them with a deeper energy-based framework.

The Illusion of Progress – How Every "Discovery" Is Just Another Fix

Over the past century, every major "discovery" in physics has been another patch, another attempt to salvage an incomplete theory.

The Higgs boson was not a breakthrough—it was a mathematical requirement to keep the Standard Model from falling apart. Quarks were not discovered—they were invented to explain the unexpected properties of protons and neutrons. Gravitational waves were "confirmed," but the interpretation was based on space-time curvature, without considering alternative explanations based on energy gradients.

Instead of genuine breakthroughs, physics keeps adjusting its models to fit observations. When galaxies didn't rotate as expected, instead of questioning gravity itself, physicists invented dark matter. When the universe's expansion didn't match predictions, rather than revisiting fundamental forces, physicists created dark energy. When the Big Bang model couldn't explain the formation of early structures, instead of rethinking cosmology, physicists added inflation.

Every time an observation contradicts a theory, physics does not ask, **"What if the theory is wrong?"** Instead, it asks, **"What can we add to make the theory work?"**

This is not science—it is mathematical patchwork. A true revolution in physics will not come from adding more particles, forces, or dimensions.

It will come from questioning the assumptions that created these contradictions in the first place.

Why the Future of Science Requires Abandoning the Patchwork Approach

For physics to move forward, it must return to first principles. Instead of adding new theoretical constructs, it must question whether the ones already accepted are real or artificial.

- Instead of treating space-time as a fabric, physics must ask: Is space actually a structured energy medium, rather than an abstract geometry?
- Instead of assuming particles are fundamental, physics must ask: Are they merely energy configurations arising from structured vacuum fields?
- Instead of forcing gravity into quantum mechanics, physics must ask: Does gravity emerge naturally from energy density variations, eliminating the need for force-carrying particles?

The structured vacuum energy model provides the missing foundation for a unified physics. It explains gravity as an energy effect rather than a distortion of space-time. It eliminates force-carrying particles by showing that forces emerge from interactions within vacuum energy. It unifies light, electromagnetism, and gravity as structured energy waves—without the need for quantum paradoxes.

This is not another patch—it is a new foundation. No more ad hoc particles. No more missing mass and energy. No more force carriers that have never been observed.

Instead of adjusting broken theories, physics must rebuild reality from the ground up. The future of science depends on abandoning the patchwork approach once and for all.

Next: The Energy-Based Universe – How Structured Vacuum Energy Eliminates Contradictions

- Why space is not empty, and why energy structure is the missing key.
- How vacuum energy naturally explains gravity, electromagnetism, and nuclear forces.
- Why mass, charge, and forces emerge from energy gradients, not particle interactions.
- The unification of physics—without extra dimensions, dark matter, or force-carrying particles.

Because the truth is simple:

Physics must stop fixing broken models and start rebuilding reality from the ground up.

13.2 The Energy-Based Universe – How Structured Vacuum Energy Eliminates Contradictions

For over a century, physics has been constrained by its own misconceptions. It has treated space as an empty void—except when mathematical models required it to have structure. It has introduced particles and force carriers—without ever explaining their true origins. It has attempted to unify conflicting theories—while ignoring the one fundamental element that connects everything: energy.

But what if the key to understanding the universe is not hidden in more particles, forces, or extra dimensions? What if the true foundation of physics is structured vacuum energy? What if mass, charge, and forces are not fundamental at all, but instead emerge naturally from energy gradients?

What if space itself is not "nothing," but an active energy medium with real, measurable effects?

This section reveals why structured vacuum energy is the missing key to solving the greatest contradictions in modern physics.

Why Space Is Not Empty, and Why Energy Structure Is the Missing Key

Physics has spent too long treating space as "empty" while simultaneously attributing real effects to it. In relativity, space-time is a mathematical construct that bends, yet has no inherent physical properties. In quantum mechanics, space is filled with fluctuating energy fields, yet these fields are treated as abstract calculations rather than tangible structures. In cosmology, space is described as "nothing," yet it somehow expands, bends, and interacts with matter.

This contradiction has led to endless problems. How can gravity act across empty space with no medium? Why does the speed of light remain constant if space is truly nothing? Why do vacuum fluctuations exist if the vacuum is supposed to be empty?

The answer is simple but revolutionary: **space is not empty—it is structured vacuum energy.** It is not a passive void but an active, dynamic medium that enables energy to propagate and interact. Every force, every motion, and every physical law is a result of interactions within this structured energy field.

This eliminates one of the greatest paradoxes in physics: how "empty space" can produce measurable effects. Gravity does not need space-time curvature—it is an energy-density effect within vacuum energy. Light does not need photons—it is an oscillation in structured energy. Quantum fields are not mathematical abstractions—they are the natural framework of the vacuum.

Once physics accepts that space is structured energy, all contradictions disappear. No more mysterious bending fabric of space-time. No more unexplained quantum fluctuations. No more missing forces or hidden dimensions. Just a universe made of energy—structured, dynamic, and complete.

How Vacuum Energy Naturally Explains Gravity, Electromagnetism, and Nuclear Forces

Modern physics treats gravity, electromagnetism, and nuclear forces as distinct, independent interactions. Gravity is supposedly the warping of space-time. Electromagnetism is described as an interaction mediated by virtual photons. The strong and weak nuclear forces require force carriers like gluons and W/Z bosons.

But if structured vacuum energy is the true foundation of reality, then forces are not separate phenomena—they are all effects of energy gradients. Gravity is not a force—it is the result of energy density variations in structured vacuum energy. Electromagnetism is not caused by virtual photons—it is the movement of charge within the vacuum field. The strong and weak nuclear forces are not independent interactions—they emerge from the way energy knots stabilize in vacuum energy.

This model eliminates the unnecessary complexity of force carriers and unifies physics at its most fundamental level. Gravity, electromagnetism, and nuclear interactions are not different forces; they are different manifestations of structured vacuum energy. They emerge naturally from the way energy distributes and interacts within space.

Once we replace force-carrying particles with structured energy gradients, physics finally makes sense. No more graviton searches. No more virtual photon exchange. No more artificial strong and weak forces. Just energy—flowing, interacting, and shaping the universe.

Why Mass, Charge, and Forces Emerge from Energy Gradients, Not Particle Interactions

Physics has long treated mass, charge, and forces as intrinsic properties of particles. Mass is assigned as an inherent quality. Charge is treated as a fundamental, unchanging trait. Forces are thought to act through discrete particle exchanges.

But these are assumptions—not discoveries. What if mass is not a fundamental property, but an effect of energy density? What if charge is not an intrinsic trait, but an energy imbalance in vacuum fields? What if forces are not transmitted by particles, but instead emerge from structured energy flow?

In structured vacuum energy, mass arises from the way energy is concentrated and stabilized within the vacuum field. Charge is the result of asymmetry in how energy distributes within the vacuum. Forces do not "push" particles—they emerge naturally from differences in energy density.

This means:

- **Gravity is not a force**—it is the natural tendency of energy to balance itself.
- **Electromagnetism is not a photon-based interaction**—it is an energy gradient effect.
- **The strong and weak nuclear forces do not require new particles**—they are local energy stabilizations within vacuum energy.

Instead of inventing new forces and particles to explain every observed phenomenon, physics must recognize that everything reduces to energy gradients in structured vacuum. Mass, charge, and forces are not separate—they are all manifestations of the same fundamental principle.

The Unification of Physics—Without Extra Dimensions, Dark Matter, or Force-Carrying Particles

For decades, physics has pursued unification through increasingly complicated theories. String theory proposed extra dimensions, yet has produced no experimental evidence. Supersymmetry created entire new families of particles, none of which have been observed. Dark matter was introduced to "fix" gravity, yet remains undetected.

All of these were distractions from the real solution. If space is structured vacuum energy, unification happens naturally. If forces emerge from energy gradients, extra particles are unnecessary. If mass and charge are effects of energy structure, their behavior is naturally explained.

This means:

- **No extra dimensions**—because everything can be described through structured energy fields.
- **No dark matter**—because galactic motion is determined by vacuum energy structure, not invisible particles.
- **No force-carrying particles**—because interactions emerge naturally from energy gradients.

Instead of chasing artificial fixes, physics must return to a single, unified principle: the universe is structured energy. Everything else—mass, charge, forces, motion—is an emergent effect.

Once we accept this, unification is no longer an impossible dream—it is the natural conclusion of physics. No unnecessary particles. No hidden forces. No mathematical illusions. Just energy—structured, flowing, and shaping the universe.

Next: Moving Beyond Space-Time – Why Energy Interactions Are the True Foundation of Reality

- Why space-time is an outdated concept that must be replaced.
- How time emerges from energy flow rather than being a separate dimension.
- Why gravity is an effect of energy distribution, not a geometric warping of space.
- The role of structured energy in defining all motion, interactions, and physical laws.

Because the truth is simple:

The universe is not made of particles and forces—it is made of structured energy.

13.3 Moving Beyond Space-Time – Why Energy Interactions Are the True Foundation of Reality

For over a century, physics has treated space-time as the fundamental framework of the universe. General relativity describes space-time as a flexible fabric that bends under the influence of mass, while quantum mechanics treats time as a background parameter, independent of physical processes. These two perspectives fundamentally contradict each other—yet physics has resisted rethinking space-time itself.

But what if space-time is not fundamental? What if it is merely an emergent effect of deeper energy interactions? What if space is not an abstract geometry, but a structured energy field? What if time is not a separate dimension, but an effect of energy flow? What if gravity is not curvature, but a direct result of energy distribution?

This section moves beyond outdated concepts of space-time and reveals why structured vacuum energy is the true foundation of reality.

Why Space-Time Is an Outdated Concept That Must Be Replaced

Einstein's theory of general relativity was one of the greatest achievements in physics, yet it also became its greatest limitation. By introducing the idea that space-time bends under mass, Einstein replaced Newton's idea of gravity as a force with a geometric model. This provided accurate predictions, but at a cost—space-time became a mathematical construct with no physical substance.

But this raises a fundamental problem: how can something with no physical existence bend? If space-time is real, why does quantum mechanics treat it as a static background? If gravity is a warping of space-time, why can't it be quantized like other forces? These contradictions suggest that space-time is not fundamental—it is an emergent effect of deeper energy interactions.

What we call "space" is actually a structured vacuum energy field. What we call "time" is simply the measurement of energy evolution. What we call "gravity" is just a gradient in this structured energy field.

Physics must abandon the outdated notion of space-time as a fundamental entity and recognize that it is a secondary effect of energy interactions. Space is not a fabric—it is a structured energy medium. Time is not a dimension—it is a consequence of energy evolution. Gravity is not curvature—it is an energy gradient.

Once physics removes space-time as a foundational concept, the contradictions between relativity and quantum mechanics disappear. The universe is not governed by abstract geometry—it is shaped by structured energy.

How Time Emerges from Energy Flow Rather Than Being a Separate Dimension

Time has always been one of physics' greatest mysteries. Newton saw time as absolute—a background clock that ticked independently of everything. Einstein showed that time was relative, changing based on motion and gravity. Quantum mechanics, however, contradicts both views by treating time as a fixed external parameter.

But what if time is not a separate entity at all? What if time is not a fundamental dimension, but an emergent effect of energy interactions? What if the "flow" of time is simply the natural evolution of energy states? What if time does not exist without energy, because without energy interactions, there is no reference for change?

In a structured vacuum energy model, time does not "flow"—it is just a measure of how energy states evolve. Time does not "slow down" in gravity—energy interactions change in response to different energy densities. Time does not exist independently—because without mass-energy interactions, there is no time reference.

There is no absolute past or future—only changing energy states. Time dilation in relativity is not an effect on "time" itself, but on how energy processes occur in different density conditions. Time appears to stop inside black holes not because time itself is disappearing, but because energy interactions reach equilibrium.

Time is not a fundamental dimension—it is simply a byproduct of energy evolution. No more time paradoxes, no more contradictions between quantum mechanics and relativity, just energy evolving naturally, without the need for an artificial time dimension.

Why Gravity Is an Effect of Energy Distribution, Not a Geometric Warping of Space

For over a century, physics has struggled to understand gravity. Newton described it as a force but could not explain its mechanism. Einstein replaced it with space-time curvature but could not explain what space-time actually is. Quantum mechanics has failed entirely to incorporate gravity into its framework.

But what if gravity is not a force at all? What if gravity is simply an effect of energy density variations in structured vacuum energy? What if it does not require force carriers or curved space-time? What if mass does not "bend space," but instead alters the surrounding energy distribution?

In this model, mass is an energy configuration that affects the surrounding vacuum energy density. Gravity is the natural tendency of energy to flow from high-density to low-density regions. Objects "fall" not because space is curved, but because energy gradients guide their motion.

There is no need for space-time curvature—gravitational effects are simply energy distribution patterns. There is no need for dark matter—galactic rotation curves can be explained by structured vacuum energy variations. There is no need for quantum gravity—because gravity is not a fundamental force, but an emergent energy effect.

By replacing curved space-time with structured energy gradients, we finally unify gravity with the rest of physics. No force carriers, no missing mass, no artificial constructs—just energy behaving as expected.

The Role of Structured Energy in Defining All Motion, Interactions, and Physical Laws

Once we accept that energy is the true foundation of reality, all physical laws become unified. Motion is not movement through "space-time"—it is the redistribution of energy within structured vacuum. Forces are not separate entities—every interaction is a result of energy gradients. Particles are not fundamental—they are stable energy configurations in the vacuum field.

Electromagnetism is not a separate force—it is an energy imbalance in vacuum fields. Gravity is not a separate force—it is an effect of energy flow. Mass and charge are not intrinsic properties—they are emergent effects of structured energy configurations.

Instead of seeing the universe as a collection of forces, particles, and dimensions, physics must recognize that the universe is a continuous energy system. All interactions are simply redistributions of structured energy. Space, time, and gravity are not fundamental—they emerge from energy flow.

Once physics adopts this perspective, unification is no longer a problem—it is the natural conclusion of understanding reality. No more separate theories for gravity, electromagnetism, and quantum mechanics. No more paradoxes between relativity and quantum mechanics. Just a single, unified framework based on structured energy.

The future of physics does not lie in refining old models—it lies in abandoning space-time altogether. Space is structured vacuum energy. Time is the evolution of energy states. Gravity is the result of energy distribution.

The universe is not a stage where objects move—it is a dynamic, self-organizing energy field.

Next: A New Scientific Revolution – From Mathematical Illusions to Physical Reality

How science must shift from abstract equations to real, testable physical principles. Why empirical validation must replace unprovable mathematical constructs. How an energy-based model opens new doors in technology, engineering, and cosmology. The next era of discovery—real science, free from the limitations of outdated theories.

Because the truth is simple: space-time was a temporary fix—structured vacuum energy is the real foundation of reality.

13.4 A New Scientific Revolution – From Mathematical Illusions to Physical Reality

For the past century, science has been trapped in an illusion. Physics has relied on mathematical constructs rather than physical explanations. Theoretical models have become more important than experimental verification. Each so-called breakthrough has been another mathematical patch rather than a fundamental discovery.

But science was never supposed to be about abstract equations—it was meant to uncover physical truth. Newton's mechanics were built on direct observations of motion. Maxwell's equations described real electromagnetic waves without unnecessary assumptions. Einstein's early work focused on explaining measurable effects, not inventing untestable dimensions.

Yet, somewhere along the way, physics abandoned reality in favor of mathematical convenience. Quantum mechanics embraced probabilities instead of asking what particles actually are. Relativity defined space-time as a geometric abstraction rather than questioning its physical existence. String theory added extra dimensions, not because they were observed, but because they made the equations work.

If science is to move forward, it must abandon this obsession with equations and return to physical reality. No more unprovable models. No more mathematical constructs that ignore physical meaning. No more theories designed to "work" rather than explain. This is the start of a new scientific revolution—one that will redefine physics on a foundation of real, testable principles.

How Science Must Shift from Abstract Equations to Real, Testable Physical Principles

Modern physics has become an equation-driven industry. The Standard Model is filled with adjustable parameters designed to fit experiments, not predict them. Dark matter and dark energy were introduced not because they were detected, but because equations didn't work without them. Quantum mechanics describes probability distributions rather than explaining what particles are.

Instead of deriving reality from math, physics must take a new approach. Start with real, testable physical principles. Build theories based on observed behavior, not just mathematical consistency. Reject models that introduce unobservable entities.

If dark matter has never been detected, it should not be assumed to exist. If extra dimensions cannot be tested, they should not be part of a physical theory. If space-time curvature does not explain the true nature of gravity, it should be abandoned.

Real science is not about making equations balance—it is about discovering what truly exists. No more hidden variables to explain missing observations. No more fine-tuning equations to fit data. No more treating physics like an abstract puzzle rather than a study of reality. The future of science must be driven by testable, observable phenomena—not theoretical convenience.

Why Empirical Validation Must Replace Unprovable Mathematical Constructs

A real scientific theory must make predictions that can be tested. Newton's laws predicted planetary motion—confirming their validity. Maxwell's equations predicted electromagnetic waves—leading to real-world applications. Einstein's relativity predicted time dilation—directly measured using atomic clocks.

But modern physics has moved away from this principle. String theory has produced no testable predictions, yet it dominates theoretical physics. Quantum gravity models rely on assumptions that cannot be observed at current energy scales. Dark energy was introduced because of equations, yet no one can directly detect it.

This is not science—it is speculation wrapped in complex mathematics. If a theory cannot be tested, it is not physics. If a phenomenon is not observed, it should not be used to modify equations. If an explanation requires unmeasurable entities, it is not a valid model of reality.

Empirical validation must be the foundation of the next era of science. Theories must be built from observed facts, not assumed mathematical structures. Experiments must guide our understanding, not be forced to fit equations. Physics must return to testable reality rather than constructing abstract models.

Only by returning to real evidence can science progress beyond its current stagnation.

How an Energy-Based Model Opens New Doors in Technology, Engineering, and Cosmology

The structured vacuum energy model is not just a theoretical shift—it is a foundation for real technological progress. By understanding energy as the true fabric of reality, we open the door to breakthroughs in energy harnessing. By recognizing structured vacuum interactions, we can revolutionize material science and electronics. By redefining gravity as an energy gradient, we can develop new propulsion and transportation technologies.

This model has immediate applications in multiple fields. Advanced energy systems can be developed by tapping into structured vacuum energy for limitless power. Electromagnetic engineering can design new wave-based devices beyond traditional particle physics limits. Gravity control can be achieved by manipulating energy gradients to revolutionize transportation.

In cosmology, this new perspective solves long-standing mysteries. Galactic rotation curves can be explained by energy distribution rather than dark matter. Cosmic expansion can be described through vacuum energy fluctuations rather than dark energy. Black holes can be understood as energy regulators rather than singularities.

Instead of chasing theoretical particles and hidden dimensions, physics must focus on the practical applications of structured energy. This is not just a philosophical shift—it is a technological revolution.

The Next Era of Discovery—Real Science, Free from the Limitations of Outdated Theories

For decades, science has been limited by its own assumptions. Physics has been trapped by the belief that space-time is fundamental. It has been constrained by the idea that particles are the building blocks of reality. It has been misguided by the obsession with force carriers, missing matter, and extra dimensions.

But the next era of discovery is already here. A shift from particles to energy configurations will redefine quantum mechanics. A shift from space-time to structured vacuum energy will revolutionize gravity. A shift from mathematical models to empirical science will bring real breakthroughs.

This is not just a revision of physics—it is the start of a new scientific revolution. One where theories are built on real, testable principles. One where experiments determine the models, not the other way around. One where science is free from mathematical illusions and grounded in physical reality.

This book has not just exposed the failures of modern physics—it has provided a path forward. No more particle zoo. No more unprovable dimensions. No more theoretical dead ends. Just one unified principle—energy as the true foundation of the universe.

This is the future of physics. A world where reality is understood, not just calculated. A world where discoveries lead to real progress, not endless speculation. A world where science is no longer a collection of patches, but a search for truth.

Final Words: The Dawn of a New Scientific Era

What happens next depends on how science chooses to move forward. Will physics cling to outdated models, or embrace structured energy as the true foundation of reality? Will the next generation of scientists continue patching old theories, or build a new framework based on testable, real principles?

The choice is ours. The path to real physics has been laid out. The universe is not a mystery—it is an energy system waiting to be understood. The time for patchwork theories is over—it is time for a true scientific revolution.

Are we ready to take the first step?

Physics Fixing: A Bold Truth for the Future of Science

Key Takeaways from the Book

- Modern physics is built on a **patchwork of mathematical fixes** rather than fundamental understanding.
- The **Standard Model** is a collection of invented particles and forces, introduced to explain inconsistencies rather than uncover deeper truths.
- **Gravity is not a force and not space-time curvature**, but an emergent effect of structured energy fields.
- **Time is not a fundamental entity**, but an illusion created by energy interactions and human perception.
- Space is **not empty**, and the vacuum is structured with energy that drives fundamental interactions.
- The **so-called "fundamental forces" (gravity, electromagnetism, nuclear forces)** are effects of deeper energy structures, not independent interactions.
- **Quantum mechanics and relativity contradict each other**, proving that our current understanding is incomplete.
- Physics must **move beyond particles, force carriers, and mathematical abstractions** to embrace an energy-based model of reality.

What If...? (Speculative Exploration)

- What if **space is not an empty void**, but an active energy system shaping motion, gravity, and electromagnetism?
- What if mass, charge, and forces **are not independent properties** but emergent effects of energy density variations?
- What if dark matter and dark energy **are not real entities**, but

misinterpretations of structured energy interactions in space?

- What if the concept of **time is just a human illusion**, and reality consists of a timeless, structured energy field?
- What if the **search for more particles** (gravitons, axions, supersymmetry) is just an endless distraction from seeking the real structure of energy?
- What if **physics abandoned force-based models** and instead studied how structured energy dictates all interactions?

Why Has Science Ignored These Issues?

- **Mathematical models took precedence over physical intuition**, making physics a game of adjusting equations rather than understanding reality.
- The **success of relativity and quantum mechanics** led physicists to ignore their contradictions rather than question their foundations.
- The **discovery of particles (Higgs, W/Z bosons, quarks)** was driven by mathematical necessity, not actual physical detection.
- **Quantum mechanics was accepted despite its paradoxes**, leading to a "shut up and calculate" approach instead of seeking deeper mechanisms.
- **The rejection of structured space (ether, vacuum energy)** forced physics into an unnatural model requiring dark matter, dark energy, and virtual particles.
- **Scientific funding and academic culture discourage questioning mainstream ideas**, leading to a cycle of reinforcing old models rather than exploring new ones.

Challenges & Open Questions

- If space is structured, **how do we measure its energy directly?**
- How does structured vacuum energy explain **why mass influences gravity?**
- Can an energy-based atomic model **explain chemical bonding**

better than quantum mechanics?**

- If forces emerge from energy gradients, **what governs their interaction at different scales?**
- If time is an illusion, **how can we reformulate physics without it?**
- How do we transition from **force-based physics to a unified energy-based theory?**

Suggested Readings & Further Exploration

- **Einstein's Own Doubts About Space-Time** – Why even Einstein questioned the completeness of general relativity.
- **Tesla's Energy-Based Model of Reality** – Why Tesla rejected space-time curvature and particle-based physics.
- **The Failure of the Standard Model** – Why modern physics still cannot explain 95% of the universe.
- **Chavan's Grand Unified Theory of Physics** – A radical rethinking of physics beyond patches, particles, and mathematical illusions.

Final Thoughts

This book has exposed the **cracks in modern physics**, dismantling the myths of force carriers, space-time curvature, fundamental particles, and the illusion of time. The next step is not just to refine equations, but to **rebuild physics from the ground up** with a focus on **structured energy, real physical mechanisms, and fundamental truths.**

The future of science is not in defending outdated theories—it is in **challenging them, questioning every assumption, and seeking a unified model that finally explains the universe as it truly is.**

Are you ready to question everything?

Epilogue

A Thought Experiment: The Great Physics Debate

(An imaginary roundtable discussion between some of history's greatest physicists, including Albert Einstein, Nikola Tesla, and Sandeep Chavan, debating the foundations of modern physics and the new structured energy paradigm.)

Moderator: "Welcome, gentlemen, to this historic discussion where we question the very fabric of reality. Today, we will debate the validity of the current physics model versus a new structured energy framework that challenges long-standing theories. Let's begin with gravity."

Question 1: Is Gravity Really the Warping of Space-Time?

◈ **Moderator:** "Einstein, you revolutionized our understanding of gravity with general relativity. Do you still believe gravity is the curvature of space-time?"

◈ **Albert Einstein:** "General relativity was the best mathematical framework I could devise to explain planetary motion and gravitational lensing. The equations work remarkably well, and experiments have confirmed my predictions. What more do we need?"

◈ **Nikola Tesla:** "Ah, Albert, always so focused on equations! But you never explained *why* mass should warp space. You simply described an effect, not a cause. Space is not an empty coordinate system that bends—it is an active energy field. Gravity is not a distortion of space-time but an emergent effect of structured energy density."

◈ **Sandeep Chavan:** "Exactly, Tesla. Space-time curvature does not provide a physical mechanism for gravity. It assumes an abstract mathematical space that can bend without defining what it's bending into. If we instead see space as a structured energy field, then gravity naturally emerges from energy gradients rather than from invisible warps in an undefined fabric."

◈ **Einstein:** "But general relativity has been experimentally verified through gravitational waves, GPS corrections, and frame-dragging. How do you account for these observations?"

◈ **Chavan:** "Those observations are real, but they do not necessarily prove that space itself is bending. They can be explained by the way mass interacts with structured energy fields. Just because your equations match reality does not mean your interpretation is correct. Newton's gravity also predicted planetary motion accurately, but we now know his interpretation was incomplete."

Question 2: Does Light Really Travel as Photons?

◈ **Moderator:** "Quantum mechanics tells us that light behaves as both a wave and a particle. Do you agree?"

◈ **Einstein:** "My work on the photoelectric effect showed that light interacts in discrete packets—photons. This particle nature is undeniable."

◈ **Tesla:** "Nonsense! Light is a continuous energy flow. There are no photons flying through space like bullets. You are mistaking energy quantization for particle behavior. The so-called 'particle' nature of light is an illusion caused by the way energy is absorbed and emitted."

◈ **Chavan:** "Tesla is correct. The photoelectric effect does not prove photons exist. It only proves that energy is absorbed in discrete amounts. This is simply how energy interactions work in structured energy fields—it does not mean light itself is made of particles. Your wave-particle duality exists only because you refuse to abandon the outdated particle framework."

◈ **Einstein:** "Then how do you explain single-photon experiments, where light appears to behave as both a wave and a particle depending on how we measure it?"

◈ **Chavan:** "What you call a 'photon' is just an energy interaction, not an independent particle. The structured energy model suggests that what you observe is the wavefront of energy disturbances interacting with the medium, not an object moving through space. Your detectors are creating the illusion of particle-like behavior."

Question 3: Why Does Quantum Mechanics Work if It's Incomplete?

◈ **Moderator:** "Quantum mechanics predicts outcomes with extreme accuracy. Why challenge it?"

◈ **Einstein:** "Quantum mechanics works, but it is fundamentally incomplete. The idea of probability waves and uncertainty disturbs me. God does not play dice with the universe!"

◈ **Chavan:** "You were right to doubt it, Einstein. Quantum mechanics works mathematically, but it is conceptually flawed. It treats atoms, electrons, and photons as probabilistic entities rather than structured energy formations. The so-called 'uncertainty' is merely an effect of structured energy interactions, not a fundamental property of reality."

◈ **Tesla:** "If physicists had focused on understanding energy fields instead of chasing particles, they would have realized that structured energy governs all interactions. The universe operates on deterministic laws, not probability waves."

Question 4: What Is the Nature of Space and Time?

◈ **Moderator:** "Is time a fundamental part of reality, or just a human concept?"

◈ **Einstein:** "Time is woven into space-time as a fourth dimension. It is inseparable from space and must be considered part of the fabric of reality."

◈ **Tesla:** "You have confused measurement with reality, Albert. Time is nothing but a way to track changes in energy states. The past and future do not exist—only the present energy interactions."

◈ **Chavan:** "Tesla is correct. Time is not an entity—it is an effect. What we call 'time dilation' is simply a result of energy interactions happening at different rates in different regions of space. If we stop treating time as an absolute dimension, we can resolve many of the paradoxes in physics."

◈ **Einstein:** "But relativity predicts time dilation, and it has been experimentally confirmed."

◈ **Chavan:** "Time dilation is real, but it does not mean time itself exists as a fundamental dimension. The structured energy field naturally explains why clocks tick at different rates without requiring time to be a physical entity."

Final Thoughts: The Future of Physics

◈ **Moderator:** "Gentlemen, before we conclude, what must be done to fix modern physics?"

◈ **Einstein:** "Physics must unify gravity and quantum mechanics. I regret that I could not complete this work in my lifetime."

◈ **Tesla:** "Stop chasing particles and accept that energy is the foundation of everything. Structured energy, not mathematical abstractions, holds the answers."

◈ **Chavan:** "We must dismantle the flawed foundations of modern physics and rebuild it using structured vacuum energy as the key to everything. Until we do, we will remain lost in a maze of particles, forces, and illusions."

Moderator's Closing Remarks:

"Today's debate reveals that while Einstein's and Tesla's ideas shaped modern science, the time has come for a new physics paradigm. Structured energy offers a way forward—one that abandons force carriers, space-time distortions, and imaginary particles. The future belongs to those who dare to question everything."

"The real question is: Are we ready to accept the truth?"

The Road Ahead

A Call for a New Scientific Revolution

For centuries, physics has been built on layers of assumptions, models, and mathematical constructs. While these frameworks have successfully predicted and described many aspects of reality, they have also created deep contradictions, paradoxes, and unanswered questions. The **patchwork approach** of adding forces, particles, and dimensions every time an inconsistency arises has led to an incomplete, fragmented understanding of the universe.

It is time for a **new scientific revolution**—one that is not bound by outdated dogmas but is driven by fundamental inquiry, open-minded exploration, and a willingness to challenge even the most sacred theories.

Breaking Free from the Patchwork

Modern physics has reached a **critical point of stagnation**:

- The **Standard Model** remains incomplete, requiring dark matter, dark energy, and extra dimensions to make sense of observations.
- **Quantum mechanics and relativity** continue to stand apart, refusing to unify despite a century of attempts.
- The **nature of gravity, light, time, and space** remains shrouded in abstraction rather than rooted in a real physical mechanism.

These contradictions **do not signal deeper truths—they expose flaws in the foundation**. Instead of covering these gaps with more patches, we must **rethink physics from the ground up**.

This is not a rejection of science—it is a demand for **a more complete, logically sound, and experimentally verifiable physics**.

The Need for New Experiments

To transition from a force-based, particle-driven model to a **structured energy framework**, new experimental approaches are required. Instead of building ever-larger colliders to search for hypothetical particles, physics must:

- **Redefine gravitational research** to test whether mass interacts with structured space instead of bending an abstract coordinate system.
- **Reinterpret light experiments** to determine whether wave-particle duality is an illusion of detection rather than a property of light itself.
- **Study vacuum energy directly**, moving beyond virtual particles and considering space as a structured energy field.
- **Develop alternative atomic models** based on energy formations rather than discrete particles.

Mainstream physics has often dismissed alternative approaches, but **real progress requires breaking away from convention**.

A Call for Interdisciplinary Collaboration

Physics has long operated in isolation, divided into separate fields with limited cross-communication. A true scientific revolution will require an **interdisciplinary approach**, bringing together:

- **Physicists** who are willing to challenge the assumptions of the Standard Model.
- **Engineers and experimentalists** who can design new ways to test structured energy models.
- **Mathematicians** who can develop an alternative formalism that describes reality without forcing artificial constructs.
- **Philosophers of science** who can guide the conceptual clarity of new models.
- **Independent researchers** who are not constrained by institutional

biases.

By merging **experimental physics, theoretical innovation, and open inquiry**, we can move beyond today's limitations and build a physics that is **coherent, unified, and deeply connected to reality**.

The Responsibility of the Next Generation

The challenge of fixing physics **cannot be left to institutions alone**. The next generation of thinkers must **question boldly, explore freely, and refuse to accept contradictions as truth**.

It is time to move away from **blind acceptance of theoretical constructs** and toward **a renewed spirit of discovery**—one that values **reality over equations, mechanisms over metaphors, and structured energy over imaginary particles**.

The road ahead is not easy, but it is necessary. The fate of physics depends on whether we **cling to old illusions or dare to seek a deeper truth**.

Are we ready to revolutionize science? The future of physics depends on it.

Rewriting the Foundations

How Physics Can Be Fixed

Modern physics has strayed from its original purpose: **to understand the fundamental nature of reality**. Instead of uncovering true mechanisms, it has become a field dominated by **mathematical formalism, imaginary particles, and forced theoretical constructs**. To move forward, physics must undergo a **complete foundational rewrite**, replacing force-based and particle-driven models with **a structured energy framework** that naturally explains the universe.

The next scientific revolution will not come from **adding more layers of complexity** but from **removing unnecessary assumptions** and discovering **a more fundamental and unified framework.**

1. Moving Beyond Force Carriers and Space-Time Curvature

One of the greatest flaws of modern physics is its **reliance on force carriers (bosons) and space-time curvature** to explain interactions. These concepts:

- Are **not directly observable**, only inferred from mathematical models.
- **Contradict each other**—quantum mechanics requires discrete interactions, while relativity treats space as a continuous fabric.
- **Fail to unify**—gravity remains separate from the quantum world because it is treated as geometry rather than an energy interaction.

A New Framework: Instead of treating forces as particle exchanges or space-time distortions, we must recognize **structured energy as the true foundation of all interactions.**

- Gravity emerges from **energy gradients** rather than warped space.

- Electromagnetism arises from **oscillations in structured energy fields**, not photon exchanges.
- Atomic stability is due to **balanced energy configurations**, not strong nuclear forces.

The laws of physics should describe **how energy organizes itself into stable, structured systems** rather than assuming interactions require mediating particles or space-time distortions.

2. Evolving Equations: A Shift to Energy-Based Interactions

Current physics equations are **patchwork solutions** that force experimental data to fit outdated models. To rewrite physics, we must:

Replace force equations with energy-density interactions

- Instead of treating gravity as an attractive force or curvature, equations should describe **energy density gradients** determining motion.
- Light should be modeled as **a continuous energy wave**, eliminating the need for wave-particle duality.

Unify interactions as emergent effects of structured energy

All fundamental forces should be derived from **energy distributions in structured space** rather than separate interactions requiring bosons.

Remove reliance on probability and virtual particles

Quantum mechanics should be reformulated based on **real deterministic energy structures**, rather than relying on probability functions.

The future equations of physics **should not be forced abstractions**—they should **reflect real physical processes** driven by structured energy.

3. New Research Areas for the Next Generation

The structured energy framework opens **entirely new fields of research** for physicists who wish to challenge mainstream thought. Some key areas include:

Understanding the True Nature of Space

- If space is a structured energy field, **how do we measure its density and variations?**
- What experiments can be conducted to test **the structured energy model vs. the vacuum concept?**

Exploring Energy-Based Gravity

- Can we detect **energy gradients around massive objects**, proving gravity is not a force?
- How do structured energy interactions explain **galactic rotation without needing dark matter?**

Reevaluating Atomic Structure

- Can we experimentally verify that **electrons are stable energy configurations**, not discrete particles?
- How do structured energy fields **replace the need for strong and weak nuclear forces?**

Developing Alternative Interpretations of Light

- How can we test whether **light is purely an energy wave rather than a photon-based phenomenon?**
- Can we demonstrate that **wave-particle duality is an illusion of measurement?**

Young physicists must **break free from institutional inertia** and push the boundaries of knowledge by questioning **not just the results but the foundational assumptions** of physics.

The Challenge Ahead: A Scientific Renaissance

Physics can no longer rely on **mathematical consistency as a substitute for real understanding.** The next revolution must be:

- **Physical, not just mathematical**—based on real mechanisms, not abstract models.
- **Unified, not fragmented**—explaining forces as effects of structured energy, not separate interactions.
- **Exploratory, not restrictive**—encouraging new ideas rather than rejecting alternative views.

We are at the tipping point of a **new scientific renaissance**—one that will dismantle **old myths and rebuild physics on a foundation of structured energy interactions.**

The real question is: **Who will lead this revolution?**

Experimental Evidence & Future Tests

For the structured energy model to gain acceptance, it must be tested and validated through **experimental evidence**. Current physics relies on **mathematical predictions**, often without direct physical verification of its assumptions. Many unexplained anomalies already suggest that **our existing models are incomplete**. By designing experiments that test **structured vacuum energy, gravity as an energy gradient, and light as pure energy flow**, we can **challenge mainstream interpretations and reveal a deeper truth about reality**.

1. Validating Structured Vacuum Energy

Modern physics **contradicts itself** by treating space as both empty (in relativity) and full of virtual particles (in quantum mechanics). If space is instead a **structured energy field**, this should be measurable.

Potential Experiments:

- **Energy Density Measurements in "Empty" Space:** Use highly sensitive detectors to measure fluctuations in vacuum energy **without assuming virtual particles exist**. Test whether vacuum energy has **structured variations** rather than a uniform background noise.
- **Casimir Effect Reanalysis:** Instead of attributing Casimir force to **virtual particles**, test whether the effect can be explained as **a pressure difference in structured vacuum energy**.
- **Modified Interferometer Tests:** Conduct Michelson-Morley-type experiments with **extreme precision** to detect structured energy variations in space.
- **Laser Experiments on Vacuum Polarization:** Test whether light propagation in a vacuum reveals **patterns consistent with a structured medium** rather than pure emptiness.

2. Testing Gravity as an Energy Gradient

If gravity is not a force or space-time curvature but instead **an emergent effect of energy density gradients**, we must look for experimental evidence beyond relativity.

Potential Experiments:

- **Mass-Induced Energy Gradient Detection:** Measure **energy density changes around massive objects** to detect whether mass interacts with an underlying energy field rather than bending space-time.
- **Drop-Tower and Free-Fall Experiments:** Conduct precision drop tests in a vacuum to measure whether **gravitational acceleration follows energy distribution patterns** rather than classical Newtonian models.
- **Reanalyzing LIGO Data:** Instead of interpreting gravitational waves as **space-time ripples**, test if they are **shockwaves in structured vacuum energy**.
- **Testing "Missing Mass" in Galaxies without Dark Matter:** Develop models where galaxy rotation curves can be explained by **energy gradients in structured vacuum energy** rather than needing dark matter.

3. Redefining Light as Pure Energy Flow

If light is a continuous energy flow rather than photons, then experiments should reveal **wave-only behavior in all contexts.**

Potential Experiments:

- **Ultra-High-Speed Double-Slit Experiments:** Capture light interactions at resolutions **high enough to see whether individual photons exist or if the "photon" illusion is a detection effect.**
- **Photoelectric Effect Reinterpretation:** Perform modified

photoelectric effect experiments that test whether **electron ejection can be explained by structured energy absorption rather than particle interactions.**

- **Vacuum Light Propagation Tests:** Test whether light in an ultra-high vacuum **exhibits variations dependent on the structure of space**, which should not happen if space is truly empty.
- **Interference Pattern Without Measurement:** Conduct experiments that prevent detector interaction to prove whether light maintains **continuous wave behavior** when undisturbed.

4. Explaining Unresolved Anomalies with Structured Energy

Several anomalies in modern physics **defy conventional explanations** but could be naturally resolved if structured energy governs interactions.

Key Anomalies to Investigate:

- **The Pioneer Anomaly:** Space probes like Pioneer 10 and 11 experienced unexplained deceleration. Could be due to **subtle interactions with structured vacuum energy** rather than an unknown force.
- **Gravitational Lensing Without Dark Matter:** Light bends more than expected around galaxies, leading to dark matter hypotheses. Could structured energy gradients explain this without needing invisible mass?
- **Galaxy Rotation Curves:** Galaxies spin faster than Newtonian gravity predicts, leading to dark matter theories. Could be explained by an **energy-based gravity model** rather than unseen particles.
- **Unexplained Quantum Entanglement Effects:** Could entanglement be an effect of **instantaneous energy communication within a structured field**, rather than "spooky action at a distance"?

- **Inconsistencies in Cosmic Microwave Background Radiation:** Temperature fluctuations in the CMB are not fully explained by the Big Bang model. Could these variations reflect **pre-existing energy structures in the vacuum?**

The Future of Physics: Moving from Theory to Experiment

The **structured energy model offers a new perspective** that can be tested through carefully designed experiments. Instead of assuming current models are correct and **forcing data to fit**, physics must be open to **alternative explanations** for known anomalies.

The future of science will be decided **not by who controls the mainstream narrative but by those who dare to test reality itself.**

Will physics continue chasing imaginary particles, or will it embrace **the structured energy revolution**?

FAQs & Common Objections

As with any paradigm shift, the structured energy model challenges long-standing theories in physics. Mainstream physicists will naturally raise objections and demand explanations for fundamental phenomena. Below, we address the most common criticisms with logical, evidence-backed responses to reinforce the credibility of this new framework.

1. "If photons don't exist, how do we explain the photoelectric effect?"

Mainstream Argument: The photoelectric effect, for which Einstein won the Nobel Prize, shows that light ejects electrons in discrete packets, proving that photons exist as individual particles.

Structured Energy Response: The photoelectric effect does not prove that photons exist—it only proves that energy is absorbed and emitted in discrete amounts.

- The process of an electron being ejected is a result of **quantized energy transfer**, not a proof of particle-like photons.
- **All waves transfer energy in discrete amounts** depending on the nature of their interaction with matter.
- The so-called "photon" is an **interpretation of quantized interactions**, not proof of an independent light particle.
- Structured energy interactions naturally explain why energy is absorbed and emitted in discrete steps without requiring a particle-like photon.

2. "If gravity is not a force, how does mass affect motion?"

Mainstream Argument: Gravity must be a force (Newtonian) or a curvature of space-time (Einstein) because mass attracts objects, causing them to accelerate. If gravity is not a force, what causes this motion?

Structured Energy Response: Gravity is an emergent effect of **energy gradients in structured space**, not a separate force.

- Objects move because they follow the **energy density landscape** of the vacuum, not because of an attractive force.
- Mass does not "pull" other objects—it **creates variations in structured vacuum energy** that determine how objects move.
- This naturally explains why gravity **affects all objects equally**, unlike electromagnetic or nuclear forces.
- The need for force-carrying gravitons disappears—gravity is simply a function of energy distribution.

3. "What evidence exists for structured vacuum energy?"

Mainstream Argument: If space is structured, we should be able to measure it. What experimental proof exists for structured vacuum energy?

Structured Energy Response: Multiple observations suggest that vacuum energy is not just a background field but **actively structured**:

- **The Casimir Effect** – This force between uncharged plates in a vacuum can be explained as a **pressure difference in structured energy** rather than virtual particles.
- **Gravitational anomalies (Pioneer anomaly, galaxy rotation curves)** – The structured energy model naturally explains these anomalies without requiring dark matter.
- **Interference patterns in quantum experiments** – Instead of invoking wave-particle duality, these can be explained by structured energy influencing motion.
- **Vacuum fluctuations in quantum field theory** – Instead of treating them as particle-antiparticle pairs, structured vacuum energy provides a more coherent explanation.

4. "Doesn't relativity prove space-time curvature?"

Mainstream Argument: General relativity has been confirmed through gravitational lensing, GPS corrections, and frame-dragging effects. How can we reject space-time curvature?

Structured Energy Response: These effects are real, but they do not require space to "bend"—they can be explained as energy interactions.

- Lensing occurs because **light follows energy gradients**, not because space itself bends.
- GPS time corrections can be explained by differences in **energy field density**, rather than treating time as a dimension.
- Frame-dragging can be interpreted as an interaction between **mass and structured vacuum energy**, rather than the rotation of an abstract fabric.

5. "How can quantum mechanics be wrong when it predicts results so accurately?"

Mainstream Argument: Quantum mechanics has been experimentally verified to extreme precision. How can it be fundamentally flawed?

Structured Energy Response: Quantum mechanics works **as a mathematical tool**, but its interpretation is incomplete.

- The equations of quantum mechanics are useful, but they **do not describe a real physical mechanism**—they only provide probabilities.
- The structured energy model removes the need for **wave-particle duality and probability functions**, replacing them with deterministic energy interactions.
- **Quantum entanglement is misunderstood**—it may be an effect of instantaneous energy communication within structured space rather than "spooky action at a distance."

6. "What about experimental proof for force-carrying particles like W, Z bosons and gluons?"

Mainstream Argument: The Standard Model has successfully predicted force-carrying particles like the W and Z bosons and gluons. If forces are just energy effects, how do we explain these results?

Structured Energy Response: These particles were introduced **as theoretical requirements to make equations work**, not because they were fundamentally necessary.

- What physicists "see" in colliders is **not direct detection of particles**, but energy interaction patterns interpreted as new particles.
- Many of these particles exist **only under extreme artificial conditions**, suggesting they are momentary energy formations, not fundamental building blocks.
- Instead of searching for **new force carriers**, we should investigate whether forces emerge naturally from energy structuring.

7. "How does your model unify all forces?"

Mainstream Argument: Physicists are still searching for a Grand Unified Theory (GUT) to unify the fundamental forces. How does structured energy solve this problem?

Structured Energy Response: The search for a "unified force theory" is misguided because **forces do not need to be unified—they need to be redefined**.

- Gravity is **not a force**, but an energy gradient effect.
- Electromagnetism emerges from **energy oscillations in structured vacuum energy**.
- The so-called strong and weak nuclear forces are **manifestations of energy stabilization in atomic structures**, not separate interactions.
- Instead of unifying four forces, structured energy shows they are **different expressions of the same underlying energy framework**.

8. "If this model is correct, why hasn't mainstream physics adopted it?"

Mainstream Argument: If structured energy is the correct approach, why haven't mainstream physicists accepted it?

Structured Energy Response: Science does not change overnight—history shows that paradigm shifts are **met with resistance until evidence becomes undeniable**.

- **Galileo was ridiculed** for claiming the Earth moved around the Sun.
- **Einstein faced skepticism** before relativity was accepted.
- The **Standard Model has been defended for decades** despite its growing contradictions.
- Change happens **when enough people start questioning**—structured energy is the next breakthrough.

Final Thoughts: A Call for Open Inquiry

Instead of **defending old models at any cost**, science must remain open to **new interpretations**. The structured energy model is not just another theory—it is **a fundamental rethinking of reality**.

To the skeptics: **Test it. Challenge it. But do not ignore it.**

Physics is not about protecting past theories—it is about **finding the truth.**

A Comparative Table – Mainstream Physics vs. Structured Energy Approach

The structured energy model challenges the long-standing assumptions of mainstream physics, offering a coherent alternative that removes the need for force carriers, space-time curvature, and abstract mathematical constructs. Below is a quick-reference comparison between the current mainstream theories and the structured energy approach.

Concept	Mainstream Physics	Structured Energy Approach
Gravity	Space-time curvature (mass bends space)	Energy density gradient (mass interacts with structured vacuum energy)
Light	Wave-particle duality (sometimes a wave, sometimes a particle)	Continuous energy flow (no photons, just structured wave interactions)
Atoms	Particle-based structure (protons, neutrons, electrons)	Structured energy formations (stable energy configurations, not discrete particles)
Forces	Exchange particles (bosons like gluons, photons, W/Z)	Emergent effects of energy fields (no need for virtual particles or force carriers)
Time	Fourth dimension of space-time	Measurement of energy interactions (not a fundamental entity)

Key Takeaways from the Comparison:

- Gravity is not a bending of space but an energy interaction.
- Light does not switch between wave and particle—it is always a structured energy wave.
- Atoms are not made of tiny particles but of stable energy formations.
- Forces do not require bosons; they emerge from structured energy interactions.
- Time is not a dimension—it is just how we track changes in energy

states.

By eliminating unnecessary assumptions and focusing on structured energy interactions, this new framework offers a unified and physically meaningful explanation of reality—one that aligns with both experimental results and logical coherence.

Open Problems & Future Research Directions

The structured energy model offers a groundbreaking perspective on physics, but like any emerging paradigm, it presents unanswered questions that must be explored through rigorous theoretical development and experimental validation. Below, we outline the major open problems and potential research areas that can guide future physicists, engineers, and interdisciplinary scientists toward a deeper understanding of reality.

1. Major Unanswered Questions

The Nature of Structured Vacuum Energy

- If space is not empty but a structured energy field, what governs its structure?
- How do energy density variations form and influence motion and interactions?
- Can structured vacuum energy explain quantum fluctuations without needing virtual particles?

Gravity as an Energy Gradient

- What physical mechanism causes mass to create energy gradients in structured space?
- Can we develop a quantitative energy-based gravity model that eliminates space-time curvature?
- How do gravitational anomalies (such as galaxy rotation curves) fit into this model without requiring dark matter?

The True Nature of Light

- If photons do not exist, how do we fully model light as a structured wave?
- What causes the apparent quantization of light interactions (such as in the photoelectric effect)?
- How does light interact with structured space, and what experiments can detect this interaction?

The Energy-Based Structure of Atoms

- If atoms are stable energy formations, not particle clusters, how do we define their internal structure?
- How do nuclear interactions emerge from structured energy rather than from strong and weak forces?
- Can an energy-based atomic model explain chemical bonding better than quantum mechanics?

The Role of Time in an Energy-Driven Universe

- If time is merely a measurement of change, how do we reformulate physics equations without it?
- What replaces time-dependent equations in relativity and quantum mechanics?
- Can we prove experimentally that time is an emergent effect, not a fundamental entity?

2. Research Topics for PhD Studies and Theoretical Exploration

For young physicists, engineers, and mathematicians, the structured energy model opens a vast field of new research directions. Below are potential topics for PhD dissertations, postdoctoral studies, and independent research.

Experimental Physics & Technology Applications

- Developing New Tests for Structured Vacuum Energy – How can we detect and measure structured energy fields?
- Reanalyzing Quantum Experiments – Can we reinterpret the double-slit experiment under structured energy physics?
- Testing Light Propagation in Vacuum – Can new laser experiments reveal deviations from mainstream wave-particle models?

Theoretical Physics & Alternative Models

- Deriving Gravity from Energy Interactions – Can we develop equations for gravity as an energy gradient rather than space-time curvature?
- Rewriting Electromagnetism Without Force-Carrying Photons – How can Maxwell's equations be modified for a pure energy-flow model?
- Developing an Energy-Based Atomic Model – What equations describe the balance of structured energy in atomic stability?

Astrophysics & Cosmology

- Explaining Galaxy Rotation Without Dark Matter – Can structured energy replace dark matter in gravitational models?
- The Role of Energy Structures in Black Holes – Can we describe black hole formation without singularities?
- Cosmic Microwave Background (CMB) Reinterpretation – Do structured energy fields explain temperature variations without inflation theory?

3. Potential Collaborations & Future Research Institutes

For the structured energy model to advance, collaboration is essential. Below are possible areas of interdisciplinary work that could bring together researchers from different domains.

Collaborating Across Scientific Disciplines

- Physics & Engineering: Developing precision instruments to measure structured vacuum energy.
- Mathematics & Computational Physics: Creating new models that replace space-time equations with energy-density interactions.
- Astrophysics & Cosmology: Testing structured energy explanations for galactic dynamics and black hole formation.
- Philosophy of Science: Re-examining the foundations of physics to support a paradigm shift.

Future Research Institutions & Independent Labs

- Encouraging universities to establish dedicated research centers for structured energy physics.
- Creating independent think tanks where researchers are free to explore non-mainstream theories.
- Partnering with technology innovators to explore practical applications of structured energy.

The Future of Physics: A Call to Action

The structured energy model is not just an alternative—it is a call for a scientific revolution. By addressing these open problems and engaging in future research, we can rewrite the foundations of physics and move beyond the limitations of today's mainstream theories.

The future belongs to those who question boldly, explore freely, and refuse to accept contradictions as truth.

Will you be part of this revolution?

Acknowledgments

This book is the result of years of deep questioning, exploration, and the relentless pursuit of truth. It would not have been possible without the support, inspiration, and contributions of many individuals who have influenced my thinking, challenged my assumptions, and encouraged me to push beyond the limits of conventional physics.

Gratitude to the Pioneers of Thought

I extend my deepest respect to the **great minds of history** who dared to question the mainstream and reimagine reality:

Nikola Tesla, for envisioning an energy-driven universe long before the world was ready to accept it.

Albert Einstein, whose brilliance reshaped physics, even as his work left behind unresolved contradictions that needed further exploration.

Isaac Newton, who gave us the foundation of mechanics but also showed that science is an ever-evolving journey.

David Bohm, for his revolutionary insights into quantum mechanics and the role of an underlying order in the universe.

All independent thinkers, past and present, who refused to accept contradictions and instead searched for deeper truths.

To My Mentors, Colleagues, and Fellow Thinkers

I am deeply grateful to those who have **engaged with me in discussions, challenged my ideas, and expanded my understanding**. Meaningful dialogue is the cornerstone of scientific progress, and I appreciate every individual who has contributed in this way.

To those who have stood by me in this intellectual journey—whether through direct discussions, academic exchanges, or shared curiosity—I thank you for **your critical insights, encouragement, and willingness to question the status quo.**

To the Scientific Community

Even as I challenge many of the assumptions in modern physics, I acknowledge and respect the vast body of work that has brought us to this point. Science is not static—it grows through debate, refinement, and the courage to admit when a model is incomplete. My hope is that this book serves as an **invitation to an open dialogue**, rather than a rejection of past achievements.

To My Readers

Your curiosity and willingness to **explore beyond established narratives** is what will shape the future of science. Whether you agree or disagree with the ideas presented here, I urge you to **continue questioning, investigating, and thinking critically**. The next great discoveries will come from those who are **unafraid to challenge conventional wisdom**.

A Final Thought

No one changes the world alone. Every shift in human understanding has been built upon the courage of those who dared to question. This book is dedicated to all those who, like me, seek to understand the true nature of reality—not by accepting the easy answers but by relentlessly pursuing the deeper ones.

With gratitude and a vision for the future,

Sandeep Chavan

Further Reading

The journey of understanding physics does not end with this book—it is only the beginning. To encourage further exploration, I have curated a list of books, research papers, and documentaries that provide deeper insights into the flaws of modern physics, alternative theories, and structured energy models. These resources will help readers question conventional narratives and explore new possibilities for the future of physics.

1. Books That Challenge Mainstream Physics

"The Evolution of Physics" – Albert Einstein & Leopold Infeld

Einstein's reflections on the changing nature of physics and the search for deeper understanding.

"Tesla: Man Out of Time" – Margaret Cheney

A deep dive into Nikola Tesla's revolutionary ideas, many of which align with structured energy theories.

"Quantum Reality: Beyond the New Physics" – Nick Herbert

A critical look at quantum mechanics and alternative interpretations that go beyond the probabilistic model.

"David Bohm: Wholeness and the Implicate Order" – David Bohm

A brilliant physicist's attempt to unify reality through an interconnected, holistic framework.

"Subquantum Kinetics: A Systems Approach to Physics and Cosmology" – Paul A. LaViolette

A theory of subquantum energy flows as an alternative to the Standard Model.

"The Trouble with Physics" – Lee Smolin

A renowned physicist explains why modern physics has reached a dead end and needs a fundamental shift.

"Einstein's Lost Key: How Relativity and Quantum Mechanics Need Each Other" – Alexander Unzicker

A critique of modern theoretical physics and the search for a unified theory.

"The Hidden Reality: Parallel Universes and the Deep Laws of the Cosmos" – Brian Greene

While still within mainstream physics, this book explores new ideas beyond the Standard Model.

"Beyond Einstein: The Cosmic Quest for the Theory of the Universe" – Michio Kaku & Jennifer Trainer

An introduction to alternative ideas in physics and unification theories.

2. Research Papers & Alternative Theories

"On the Electrodynamics of Moving Bodies" – Albert Einstein (1905)

Einstein's original special relativity paper—useful for understanding where current physics began.

"A Structured Energy Approach to the Universe" – Sandeep Chavan

A detailed explanation of the structured energy model and how it challenges modern physics.

"Quantum Field Theory: A Philosophical Analysis" – David Wallace

A deep critique of quantum field theory and its inconsistencies.

"The Energy Wave Theory: A Unification of Physics" – Jeff Yee

A lesser-known but compelling attempt to unify fundamental forces without force carriers.

"Pilot-Wave Theory and Quantum Mechanics" – Louis de Broglie & David Bohm

A deterministic alternative to quantum probability that aligns with structured energy fields.

"Structured Vacuum Energy as a Replacement for Dark Matter" – Independent Research Studies

A growing body of work suggests that dark matter is an illusion created by a misunderstanding of vacuum energy.

3. Documentaries & Video Lectures

"The Principle" (2014)

Challenges the Copernican Principle and explores alternative ideas in cosmology.

"Einstein's Unfinished Symphony" (PBS)

Discusses Einstein's struggle to unify physics and the unanswered questions left behind.

"Nikola Tesla: The Genius Who Lit the World"

A must-watch documentary on Tesla's vision of an energy-based universe.

"The Simulation Hypothesis"

While controversial, this film questions the nature of reality and whether energy structures define everything.

"The Fabric of the Cosmos" – Brian Greene (PBS)

Explores space, time, and the limits of modern physics.

"David Bohm: The Quantum World Beyond the Mainstream"

Explains Bohm's alternative view of quantum mechanics and its implications.

4. Online Research & Communities

- The Foundational Questions Institute (FQXi) – A research group exploring alternative theories in physics.
- arXiv.org – A great source for cutting-edge papers on alternative physics, cosmology, and quantum mechanics.
- Independent Research Communities – Websites and forums where physicists discuss non-mainstream models.

A Final Note

These books, papers, and documentaries are not meant to be final answers—they are meant to expand perspectives and encourage questioning. Science moves forward when people dare to challenge assumptions, explore new models, and push the boundaries of knowledge.

If you seek the truth, you must read widely, question deeply, and never stop exploring.

About the Author

Sandeep Chavan

Sandeep Chavan is an independent researcher, educator, and engineer dedicated to challenging the foundations of modern physics and redefining scientific thought. With a background in mechanical engineering and industrial systems, Chavan brings a unique interdisciplinary approach to understanding the flaws in mainstream physics and proposing a structured energy-based model as a more coherent and unified theory of reality.

Throughout his career, Chavan has been driven by a deep intellectual curiosity and a relentless pursuit of truth. His research critically examines space-time curvature, force carriers, and particle-based physics, arguing that physics has become a patchwork of mathematical fixes rather than a true reflection of nature. His work presents structured vacuum energy as the fundamental medium that governs all interactions—eliminating the need for imaginary forces, dark matter, and wave-particle duality.

As an educator, Chavan has inspired countless students and researchers to think beyond conventional narratives and challenge scientific dogma. His mission is to encourage open scientific inquiry, where ideas are tested not based on their popularity but on their ability to explain reality without contradictions.

Chavan's research spans a wide range of topics, including:

- The fundamental flaws in quantum mechanics and relativity
- An alternative approach to gravity as an energy density gradient
- The redefinition of light as pure energy flow, not photons
- The structured vacuum energy model as the key to unification
- How time is an illusion and not a fundamental dimension

His contributions to theoretical physics and his commitment to questioning the status quo make him a leading voice in the movement toward a new scientific revolution.

When he's not engaged in research and writing, Chavan continues to explore new perspectives in philosophy, human behavior, and technology, blending science with a broader understanding of knowledge and existence.

This book, *Physics Fixing: A Bold Truth for the Future of Science*, is not just a critique of modern physics—it is a call to action for all thinkers, physicists, and researchers to question, explore, and seek the real fundamental principles that govern our universe.

Research Papers:

- ***Chavan's Grand Unified Theory of Physics:*** https://doi.org/10.5281/zenodo.14791053
- ***The Myth of Wave-Particle Duality:*** https://doi.org/10.5281/zenodo.14793162
- ***Formation of Black Holes:*** https://doi.org/10.5281/zenodo.14788539
- ***Chavan's Law: A Universal Energy Equilibrium Model for Black Holes:*** https://doi.org/10.5281/zenodo.14782900

Disclaimer

The content of this book, *Physics Fixing: A Bold Truth for the Future of Science*, is intended to provoke critical thought, encourage scientific debate, and challenge existing paradigms in physics. The ideas presented here offer an alternative framework to modern physics and are based on independent research, theoretical exploration, and logical analysis.

This book is not intended to serve as an official academic textbook or as a replacement for mainstream physics education. The structured energy model and its implications are meant to inspire further investigation, discussion, and open inquiry among physicists, researchers, students, and anyone curious about the fundamental nature of reality.

The author acknowledges that science is an evolving discipline, and while this book critiques widely accepted theories such as general relativity, quantum mechanics, and the Standard Model, it does so with the aim of encouraging deeper analysis and potential improvements to our understanding of physics. Readers are encouraged to explore both mainstream and alternative perspectives and engage in their own independent research and critical thinking.

Furthermore, while the book questions established scientific theories, it does not dismiss the valuable contributions of historical and contemporary physicists. The purpose is not to reject science but to drive it forward by questioning its foundations and assumptions.

Any references to historical figures, scientific institutions, or research studies are provided for discussion and analytical purposes only. The author does not claim to represent or speak on behalf of any academic organization, research institution, or professional body.

The structured energy model proposed in this book is an ongoing subject of exploration, and its validity must be tested through rigorous scientific methodology and experimental verification. As with any new scientific perspective, further research, peer review, and open discourse are essential to its evaluation.

The author assumes no responsibility for any misinterpretation, misuse, or misapplication of the information presented in this book. Readers are advised to apply their own judgment and seek guidance from multiple scientific sources when forming conclusions.

Don't miss out!

Visit the website below and you can sign up to receive emails whenever SANDEEP CHAVAN publishes a new book. There's no charge and no obligation.

https://books2read.com/r/B-A-EPGPC-SNLYF

Connecting independent readers to independent writers.

Did you love *Physics Fixing*? Then you should read *Never Trust Without Doubt*[1] by SANDEEP CHAVAN!

[2]

In ***Never Trust Without Doubt: The Interplay Between Doubt and Trust in an Unethical World***, **Sandeep Chavan** explores the delicate balance between trust and doubt in an era where ethical behavior is often uncertain. This thought-provoking book challenges conventional notions of trust as a blind leap of faith, instead advocating for a thoughtful approach where doubt acts as a guiding ally. It provides readers with a practical and philosophical framework to navigate relationships and decisions in a complex, morally ambiguous world.

The book delves into the mechanics of trust and doubt, redefining trust as a flexible, intentional process that evolves over time. Through chapters like Trust Reimagined and Building Trust with Boundaries, it emphasizes the importance of setting clear expectations, establishing boundaries, and using doubt as a filter to ensure trust is grounded in reality. Chavan introduces

---

1. https://books2read.com/u/m212B1

2. https://books2read.com/u/m212B1

tools such as Trust Tests, Conditional Trust Agreements, and Doubt Verification Checks, offering actionable strategies for building trust gradually and responsibly.

A key focus of the book is its recognition of doubt not as a hindrance but as an essential partner in trust. Doubt prompts reflection, encourages verification, and fosters accountability, making trust more resilient and meaningful. Through real-life case studies and practical exercises, the book demonstrates how doubt has been instrumental in helping individuals and leaders make ethical decisions, even in challenging circumstances.

Chavan also addresses the emotional dimensions of trust, highlighting the role of emotional resilience in managing trust breaches and fostering recovery without cynicism. Readers are guided on how to balance openness with self-protection, enabling them to build connections that are authentic yet secure. The book concludes with an invitation to develop a personal trust philosophy—an intentional mindset that integrates trust and doubt to create relationships and decisions aligned with one's values.

With its blend of philosophical insight, practical tools, and real-world relevance, Never Trust Without Doubt offers a fresh perspective on trust for readers navigating today's ethically complex world. It empowers individuals to embrace doubt as a catalyst for thoughtful decision-making, cultivating trust that is both resilient and transformative. This book is an essential guide for anyone seeking to build meaningful connections without compromising their integrity or well-being.

Read more at https://www.sandeepjchavan.com/.

Also by SANDEEP CHAVAN

Echoes of War: The Russia-Ukraine Conflict and Its Global Ripples

Shattered Lives

The War Economy

Borders Redrawn

Arms and Allies

Collateral Damage

Justice in the Ashes

Somewhere in the Darkness of the Cosmos

The Blueprint of the Universe

The Cosmic Breath

The Eternal Sage Chronicles

The Sage and the Sovereign

The Family Pillar

The People's Flame

The Warrior's Soul

The Fundamental Theory of Life & Existence

Energy in Action: The Soul Demystified

The Active Mind

Energy in Connection
Death: The Disconnection
Subconsciousness: The Chosen Disconnection

Ultra Modern Life Philosophy
The Duality of Laughter and Tears
Compare to Accept

Standalone
It's Not AI, It's AHI - Amplified Human Intelligence
The Decision Paradox
The Four Sapiens
Malicious Script of Indian Polity
The IIT Legacy & Global Impact
The IIT Legacy & Global Impact
You are just a Version of Your Original
Win the Game You Didn't Choose
Always Keep Your Bags Packed
Reveal to Shield - Learn the Game of Facades
Engage Beyond Elections
Culture, Identity & Change: The Evolution of Indian Society
The Triangular Dynamics
Never Trust Without Doubt
If You Can't Change the Country, Leave the Country
Live Fearless with Fear
India: The Investment Magnet
Realignment
Powering a Billion Dreams
Universal Skill
Extension: From Brush to Algorithm
Human Side of Einstein
The Future Knows No Lines

Let's Bring the Generations to the Table
Unified Energy Dynamics
A Beat That Never Dies
Strategic Horizons
Right From Scratch
The Forgotten Innovators
Physics Fixing

Watch for more at https://www.sandeepjchavan.com/.

www.ingramcontent.com/pod-product-compliance
Lightning Source LLC
LaVergne TN
LVHW090307210926
848480LV00001B/162

* 9 7 9 8 2 2 7 4 5 8 3 2 2 *